A People's History of Computing in the United States

A People's History of Computing in the United States

Joy Lisi Rankin

Cambridge, Massachusetts
London, England *2018*

Printed in the United States of America

First printing

Library of Congress Cataloging-in-Publication Data

Names: Rankin, Joy Lisi, 1976– author.
Title: A people's history of computing in the United States / Joy Lisi Rankin.
Description: Cambridge, Massachusetts : Harvard University Press, 2018. | Includes bibliographical references and index.
Identifiers: LCCN 2018009562 | ISBN 9780674970977 (alk. paper)
Subjects: LCSH: Computer systems—United States—History—20th century. | Computer networks—United States—History—20th century. | Information commons—United States—History—20th century.
Classification: LCC QA76.17 .R365 2018 | DDC 004.0973—dc23
LC record available at https://lccn.loc.gov/2018009562

For Scott and Lucy

Contents

A People's History of Computing in the United States

Introduction

People Computing (Not the Silicon Valley Mythology)

The students at South Portland High School buzzed with enthusiasm; the wires in their classroom walls hummed with information. Young men and women played games on their computing network—tic-tac-toe and checkers, solitaire and bridge, basketball and bowling. They clamored for news from other schools on the network, which crisscrossed New England and connected rural Maine to suburban Connecticut. "Don't forget to sign me up for time," they reminded each other.[1] Some of these notoriously difficult-to-rouse high schoolers even rolled out of bed at four in the morning for network access.[2]

A thousand miles to the west, students and teachers in the Minneapolis suburbs spoke the same language. From the basic building blocks of commands including IF, THEN, LET, and PRINT, they created music and poetry and solved math problems. In a few years, three of them (both students and teachers) would invent the beloved game *The Oregon Trail.*

Some four hundred miles to the southeast, engineers at the University of Illinois at Urbana-Champaign perfected a graphical plasma display screen for their computing system, a network that would soon connect people across the United States with communications options we recognize today as instant messaging and screen sharing.

Public intellectuals called for computing as a public utility, comparable to electricity or water.

The year was 1968.

When I began researching this book, dozens of vignettes like these appeared on the pages of newsletters, grant proposals, research reports, and newspaper and journal articles, and I was stunned. The people in these histories, the geography of their networks, and even the dates of their activities appeared at odds with both conventional histories of computing and powerful popular narratives.

The origin stories around contemporary American digital culture—our 24/7 connected, networked, WiFi, smartphone, tablet, Instagram, Facebook, Tweeting, thumbs-up/thumbs-down world—center on what I call the Silicon Valley mythology. This compelling myth tells us that, once upon a time, modern computers were big (and maybe even bad) mainframes. International Business Machines, much more familiar as IBM, dominated the era when computers were the remote and room-size machines of the military-industrial complex. Then, around 1975, along came the California hobbyists who created personal computers and liberated us from the monolithic mainframes. They were young men in the greater San Francisco Bay Area, and they tinkered in their garages. They started companies: Steve Jobs and Steve Wozniak established Apple; Bill Gates and Paul Allen developed Microsoft. Then, in the 1990s, along came the Internet to connect all of those personal computers, and the people using them. Another round of eccentric nerds (still all young white men)—Jeff Bezos, Sergey Brin, Larry Page, and Mark Zuckerberg among them—gave us Amazon, Google, Facebook, and the fiefdoms of Silicon Valley. Walter Isaacson's *The Innovators* expands the popular narrative of digital history to include less familiar contributors such as the nineteenth-century mathematician Charles Babbage and the twentieth-century computing visionary J. C. R. Licklider. However, Isaacson, like many others, still portrays technology as the realm of engineers, experts, and inventors, or, as his subtitle declares, "hackers, geniuses, and geeks." Computer technology, in this mythology, is far removed from everyday life until it reaches the users.

Historians have certainly complicated this Silicon Valley mythology.[3] They have pointed out the fundamental roles of federal government funding and university research in producing American computing, and they have highlighted computing's cultural origins in the Cold War and the counterculture.[4] They have taken their investigations well beyond the Bay Area, spotlighting the contributions of Boston, Tysons Corner (Virginia), and Minnesota.[5] Recent scholarship has explored computing in China, England, France, India, and the Soviet Union.[6] Scholars have produced excellent business histories and excellent histories of the people whose work entailed computing.[7] But by and large, historians have assumed that personal experiences of digitization began with the emergence of personal computers in the late 1970s, and that experiences of social computing commenced with the popularization of the Internet in the late 1990s.

The Silicon Valley mythology does us a disservice. It creates a digital America dependent on the work of a handful of male tech geniuses. It deletes the work of the many individuals who had been computing, and it effaces their diversity. It masks the importance of the federal government as a principal financial investor in digital development during the 1960s and 1970s. It minimizes the roles of primary and high schools, as well as colleges and universities, as sites of technological innovation during those decades. The Silicon Valley story is neat and pat, but it prevents us from asking how digital culture truly evolved in the United States. In short, this mythology misses the story at the heart of the transformation of American culture during the past fifty years.

The people in *A People's History of Computing in the United States* are the students and educators who built and used academic computing networks, then known as time-sharing systems, during the 1960s and 1970s. Time-sharing was a form of networked computing in which multiple computing terminals were connected to a central computer via telephone lines. It was called time-sharing not because one user had an allotment of computing time, and then another user had another allotment of computing time, but because the computer was programmed to monitor—and allocate—its own processing time among multiple simultaneous users. Multiple users could work on their individual terminals, which I identify as personal terminals, simultaneously. Terminals were located in such social settings as middle school classrooms, college dorm rooms, and university computing labs. Because the terminals relayed information to and from the central computer by telephone line, terminals could be—and were—located hundreds of miles away from the processing computer.

Make no mistake, these were networks.[8] Any user could communicate with the central computer and with another user at another location on the system via the central computer. As will be seen in subsequent chapters, students and educators embraced the computing and communications dimensions of their time-sharing networks. The possibility of storage on a central computer meant that users could share useful and enjoyable programs across the network. For example, users on the New England network, based at Dartmouth College, produced and used multiplayer games and, in 1968, a program called MAILBOX for sending messages over the network.[9]

A People's History of Computing in the United States focuses on the users of these time-sharing networks to develop a history of the digital age that emphasizes creativity, collaboration, and community. Time-sharing networks emerged neither from individual genius nor from the military-industrial complex; rather, they were created for—and by—students and educators at universities and public schools as civilian, civic-minded projects. At their most idealistic, the developers of these systems viewed access to computing as a public good available to all members of a collective body, whether that body consisted of a university, a school system, a state, or even a country.

For the students and educators, sharing was a feature, not a bug, of the networks. By design, time-sharing networks accommodated multiple users, and multiple users meant possibilities for cooperation, inspiration, community, and communication. Personal computer purveyors and boosters later insisted on the superiority of personal machines. They celebrated not having to share a computer; rather, they praised the individual access of one person to one computer. Ultimately, in the Silicon Valley mythology, the personal computer became the hero, the liberator that freed users from the tyranny of the mainframe and the crush of corporate IBM. Yet time-sharing users benefited from their technological and social networks. The computing contemporaries with whom they could exchange ideas, programs, tips, and tricks became an exceptional human resource.

The actors and networks in *A People's History of Computing in the United States* are new to American technological narratives, and so are their geographical and educational contexts. I showcase the contributions of K–12 and liberal arts college classrooms, as well as education-focused university research labs, as key sites of innovation during the 1960s and 1970s.[10] I examine the Dartmouth Time-Sharing System, which stretched across and beyond New England, the educational networks in Minnesota that culminated in statewide computing with the Minnesota Educational Computing Consortium, and the University of Illinois PLATO (Programmed Logic for Automatic Teaching Operations) System. These were not the digital cultures of Silicon Valley. Usually we think of public schools and college classrooms as the last stop for mature technology. But in the story told here, I open up a digital world in which innovation was not limited to garage hobbyists, eccentric entrepreneurs, or military-funded scientists.

I introduce the concept of "computing citizens" to describe those who accessed time-sharing networks. In this *People's History of Computing in the United States*, the definition of a computing citizen hinges on membership in a computing community. This is a broad and inclusive definition of citizenship, mirroring the ways in which the advocates of time-sharing networks envisioned computing access as broad and inclusive. Here, too, citizenship emphasizes the communal institutions, such as schools, universities, state governments, and the National Science Foundation, that enabled access and participation. I chose the term "computing citizens" to be more encompassing than producers or makers, and to differentiate them from users.[11] "User" is now synonymous with "end user" or "consumer." But in many cases, the computing citizens were not merely end users or consumers.[12] They produced and engaged in personal and social computing.[13] They built these time-sharing networks. They wrote programs for problem solving, personal productivity, and creative expression. They computed art and poetry and music. They developed methods to bank and share their programs, and they communicated by computer. Students and educators constructed networks: the technical connections among terminals and computers and telephone wires, but more importantly, the social and sociable interpersonal networks. They formed communities around their zeal for computing. "Computing citizens" simultaneously conveys their individual choices, actions, and activities, and their collective access to a social, communal resource.

This book's characterization of computing citizens is not explicitly intended in the political sense of citizenship, but questions of political membership can nevertheless be explored through the makeup of each network.[14] For instance, the time-sharing network based at Dartmouth College originated in part because college administrators viewed their students as future business, intellectual, and political leaders of the United States, and those administrators deemed computing experience essential to their leadership preparation. Although the PLATO network based at the University of Illinois started as an experiment in education, its citizens devised explicitly political uses, such as producing a program about an environmental issue. The state of Minnesota, a high-technology hub during these decades, enacted communal and political computing citizenship by creating a statewide time-sharing network for all public school students, from K–12 to community college and university.

Chapter 1 shows how college students—and college users more generally—were central to the creation of the time-sharing network based at Dartmouth College, in Hanover, New Hampshire. In the early 1960s, the mathematics professors Thomas Kurtz and John Kemeny elevated user convenience in the design of their time-sharing system at Dartmouth. Their commitment to simplicity of use, instead of efficiency for the computer, combined with their commitment to free computing for all students, set them apart from the academic, industrial, and military computing mainstream.

Chapter 2 demonstrates how the campus context, with its focus on football and fraternities, shaped the development of masculine computing at Dartmouth. I am not the first historian to argue that computing became increasingly masculine during the 1960s and 1970s; however, Dartmouth's rich archival records enable an in-depth study of the interplay of gender and computing among both computing employees and casual network enthusiasts. Studying the roles and representations of the women employed at Dartmouth's Computation Center illuminates how the gender roles of the Cold War nuclear family informed college computing.

Chapter 3 argues that BASIC (Beginners' All-purpose Symbolic Instruction Code), the programming language created for the Dartmouth network, became the language of computing citizens during the 1960s and 1970s. By 1968, students at twenty-seven New England secondary schools and colleges practiced BASIC on their terminals, connected to each other and to Dartmouth via the time-sharing network. BASIC proved central to the growth of personal and social computing, from New England westward to Minnesota, and ultimately to northern California. This chapter also examines how BASIC spread via the connected efforts of educational computing enthusiast Bob Albrecht and his *People's Computer Company* newsletters; the Huntington Project, which produced wildly popular educational computer simulations in BASIC; and the Digital Equipment Corporation, which distributed Huntington Project materials at low or no cost to sell its BASIC-enabled minicomputers to schools across the United States. Chapter 3 underscores the creativity of the BASIC citizens of the early digital era.

In the computing mythos, Americans benefited from an inexorable march from mainframes to minicomputers to microcomputers—otherwise known as PCs. Those who have complicated that story have presented time-sharing as a short-lived phenomenon of the 1960s, and they have focused on three things: the Massachusetts Institute of Technology (MIT), its Multics time-sharing project, and the financial market for time-sharing. They have overlooked that time-sharing systems were networks and that users appreciated the communication and information-processing capacities of those networks. They have also overlooked the vision of computing for the public good that emerged with time-sharing—the vision for community computing utilities.

Chapter 4 unearths the extensive discourse about computing as a utility comparable to electricity, telephone, or water service, and it highlights the numerous computing networks that emerged between 1965 and 1975. This chapter argues that widespread computing citizenship via computing utilities seemed far more promising in the 1960s and 1970s than the Silicon Valley mythology would have us believe. Neither time-sharing nor the vision for networked computing for the public good were short-lived, nor did they exclusively parallel the MIT-Multics-markets trajectory that others have highlighted.

Chapter 5 exemplifies the unifying themes of the previous four chapters. It analyzes the drive for, and development of, a statewide public computing utility. It delineates how computing citizens implemented a time-sharing network, and it highlights how BASIC enabled their personal and social computing. From 1965 to 1980, Minnesota led the nation in creating computing citizens by implementing statewide interactive computing at its public schools and universities, reaching hundreds of thousands of students. The students and educators in computing collectives, including Total Information for Educational Systems (TIES) and the Minnesota Educational Computing Consortium (MECC), developed new modes of software sharing, software banking, and software translation. By the late 1970s, Minnesota students played games such as their beloved *Oregon Trail* thousands of times every month. TIES and MECC illustrate a radically alternative vision of networked computing.

Chapters 6 and 7 move away from public schools and small colleges to a more typical setting for technological development, a large research university. But here, too, education engendered the network. During the 1960s at the University of Illinois, Donald Bitzer recruited and united students and scholars from multiple disciplines to support the creation of a personal computing terminal for education, described in Chapter 6. Bitzer and his colleagues initially created their PLATO system to explore the potential uses of computing in education, but Bitzer's drive to expand the project motivated him to open it to users across and beyond the Urbana-Champaign campus. PLATO began as a rudimentary time-sharing system, but Bitzer's emphasis on usability propelled the development of revolutionary personal terminals featuring flat-panel plasma display screens and touch-responsive screens, connected in a vast social network—all before 1975.

The 1960s and 1970s were a crucible for contemporary culture, and PLATO users developed practices that are now integral to our modern digital experience. In Chapter 7, I argue that PLATO's distinctive and evolving personal terminal, together with Bitzer's ongoing efforts to create as many PLATO users as possible, fostered a rich social network, partially funded by the Advanced Research Projects Agency (whose better-known investment was ARPANET, which became a foundation of the Internet). By 1975, the 950 terminals on the nationwide PLATO network enabled "on-line" communication in the form of bulletin boards, instant messages, and electronic mail. PLATO users swapped jokes and stories every day on their online network, and they reveled in this new sociability. At the same time, they struggled with security, censorship, and harassment, and their interactions revealed a gendered digital divide.

The Epilogue emphasizes the significance of the myriad connections among the students, educators, communities, and corporations in *A People's History of Computing in the United States.* I contend that each of the computing communities described in previous chapters struggled with the transition from computing citizenship to computing consumption. PLATO's revolutionary plasma screens attracted the investment of the Control Data Corporation, which tried (unsuccessfully) to market its own version of the PLATO system to schools and universities. The BASIC programs shared freely around the Dartmouth network and on the pages of the *People's Computer Company* newsletter fueled the imaginations of many—including Steve Wozniak and Bill Gates. Gates first learned to program in BASIC, the language on which he built his Microsoft empire. Wozniak adapted Tiny BASIC into Integer BASIC to program his homemade computer, the computer that attracted the partnership of Steve Jobs and launched Apple. And the Minnesota software library, mostly BASIC programs including *The Oregon Trail,* proved to be the ideal complement for the hardware of Apple Computers. During the 1980s, the combination of Apple hardware and MECC software cemented the transformation from computing citizens to computing consumers.

The title of this work nods to Howard Zinn's groundbreaking *A People's History of the United States.* Published nearly forty years ago, Zinn's book channeled the energies of the social and political movements of the long 1960s and the ensuing outpouring of social history to write a new kind of American history. Zinn did not write about Founding Fathers and presidents, captains of industry, war heroes, and other influential white men. Instead, he featured people rarely seen or heard in synthetic or textbook history to that point, including Cherokee and Arawak Native Americans, young women factory workers, enslaved African Americans, socialists, and pacifists.

The history of computing and networking has likewise been dominated by a Great White Men (and now, maybe a handful of women) storyline. Part of the Silicon Valley mythology is that the Information Age had Founding Fathers, men including Jobs, Gates, and Zuckerberg. According to this origin story, there were no computers for ordinary people—no personal computing—until those Founding Fathers and their hardware and software made computing accessible to everyone. Business and government leaders around the world look to Silicon Valley for guidance, inspiration, and emulation, but the Silicon Valley ideal venerates grand men with grand ideas. That narrative, by focusing on the few, has obliterated the history of the many: the many people across the United States and around the world who have been computing in different ways for decades.

This is a people's history of computing because it tells the story of hundreds of thousands of computing citizens. Like Zinn's history from the bottom up, this is a history from the user up. *A People's History of Computing in the United States* demonstrates how people experienced and shaped computing and networking when it was not central to their employment responsibilities. I identify it as a people's history to differentiate it from the Silicon Valley stories. This is not a history of great white men, or even a history of small teams of innovators. Certainly, in comparison with Zinn's actors, the people in this book could be considered elite, in that they were affiliated with educational institutions, and the people at those educational institutions were predominantly white. Moreover, as the following chapters demonstrate, men were more likely to have their computing citizenship recognized than women. Nonetheless, the students, teachers, and professors who populate this book constitute a critical group whose contributions have been overlooked in American computing history.

We have lost our computing citizenship. We consume computing via ubiquitous laptops, smartphones, and tablets. The sharing we do now is asymmetrical; we divulge the intimacies of our daily lives for the products of social media, and for the conveniences of on-demand watching, shopping, and searching. These concessions are neither collaborative nor communal. The corporations that dominate digital culture are, after all, profit driven. They increasingly act with the powers of governments, but without the responsibilities and protections that legitimate governments owe their citizens. Even the notion of net neutrality as a public good is under threat by regulation that empowers corporations at the expense of users. Although Internet access—computing access—is increasingly recognized as a necessity around the world, it is no longer conceived as a civic project.

We need histories not of computers but of the act of computing. *A People's History of Computing in the United States* spotlights how the computing of 1960s and 1970s students and educators inaugurated America's network society. It highlights the centrality of education—at all levels—as a site of creativity, collaboration, and innovation. This book showcases the benefits of national investment in education and research, as well as the crucial role of local and state governments in supporting those endeavors. We are digital consumers now. This is a history to inform and inspire the global digital citizenry we may yet become.

1

When Students Taught the Computer

In 1958, Tom Kurtz wanted to run a computer program. He woke early on a Tuesday morning and drove five or so miles from his home in Hanover, New Hampshire, to the train station in White River Junction, Vermont. He brought with him a steel box. At the station, Kurtz boarded the 6:20 train to Boston and settled in for the three-hour ride, during which he would read to pass the time. On his arrival in Boston, he took a cab to MIT's campus in Cambridge. Finally reaching the computer center at MIT, he opened the steel box. It contained hundreds of cardboard cards measuring about three inches by eight inches. One set of those cards, precisely ordered and held together with a rubber band, constituted his computer program. Other sets were programs created by colleagues at Dartmouth College, where he was a professor in the mathematics department. It was thanks to Dartmouth's participation in the New England Computation Center at MIT that they had access to an IBM 704 mainframe computer. After Kurtz handed the stacks of cards over to an employee at the center he had several hours to wait. On some occasions when he made this trip to Cambridge, he met with colleagues at MIT or nearby Harvard; other times he simply strolled around the city. Late in the afternoon, he returned to the computer center to pick up the cards, along with the precious printouts of each program's results. Reviewing them on the evening train back to White River Junction, Kurtz saw that the results for his program runs contained error reports—yet again. Finally back at home in Hanover at the end of a long day, he was already thinking of how he might revise his program in the coming days, replace some cards with newly punched ones, and go through the process all over again two weeks later.[1]

A decade later, in 1968, Greg Dobbs, a student at Dartmouth College, wanted to run a computer program. He stepped out of his dormitory, Butterfield Hall, and walked a few hundred yards north to Webster Avenue, enjoying the September sunshine. He turned right on Webster and walked just a block to the new Kiewit Computation Center. At night, he could see Kiewit's lights from his dorm room window. As he made his way to one of the few empty teletype terminals, he recognized some of his friends and classmates among the thirty or so students sitting at teletypewriters. He went through the habitual steps of the login routine, beginning by typing HELLO and pressing RETURN, and settled in to a game of FOOTBALL against the computer, typing his commands and receiving responses within seconds. He, like 80 percent of his student peers and 40 percent of Dartmouth faculty, embraced this new personal and social computing.[2]

In the early 1960s, computers were remote, inaccessible, and unfamiliar to Americans. The approximately six thousand computer installations around the nation clustered in military, business, and research-focused university settings. Individual access to computing in 1958 had been so rare, and so valuable, that Kurtz was willing to devote an entire day to gain the benefit of a few minutes of it. Within a decade, however, Kurtz and his colleague John Kemeny, together with a group of their students at Dartmouth, had transformed computing by creating an interactive network that all students and faculty, not just those working in the sciences or engineering, could use. This chapter argues that Kurtz, Kemeny, and their student assistants put the user first in the design and implementation of their network, thereby creating computing for the people. Their focus on simplicity for the user, instead of efficiency for the computer, combined with their commitment to accessible computing for the whole student body, set them apart from the mainstream of academic, industrial, and military computing.

The Problems with Mainframes

Computers were far from quotidian in 1958. In the Cold War context of the 1950s, the American military developed computing for defense against the Soviet Union with projects such as the extensive Semi-Automatic Ground Environment (SAGE) system to protect against Russian airborne attacks. Less than a year after the Soviet Union's 1957 launch of its *Sputnik* satellite alarmed Americans, President Dwight Eisenhower requested from Congress a staggering $1.37 billion "to speed missile development and expand air defenses," of which $29 million was for SAGE.[3]

This news conveyed that computers were essential to American protection—powerful and significant, but also remote and intimidating. During this post–World War II decade, American businesses ramped up both their production and their usage of computers. Remington Rand installed some of the earliest electronic, digital computers sold commercially in the United States—at the Census Bureau in 1951 and at General Electric (GE) in 1954. During that time, IBM competed with Remington Rand for leadership in the computer manufacturing field, but together they had only nine installations by the end of 1953.[4] Although computers proliferated in military, commercial, and university spaces—with several thousand in use by 1960—they functioned behind the scenes. They were used, for example, to maintain consistent oil output at Texaco's refinery in Port Arthur, Texas; to process checks for Bank of America; and to manage orders and inventories for Bethlehem Steel. In short, computers remained invisible to most Americans. Even when Kurtz visited the MIT Computation Center, he did not interact with the computer there.

Kurtz's MIT experience was emblematic of programming in the era of mainframe computers. These machines were large and therefore demanded large spaces. The IBM 704 Data Processing System Kurtz used at MIT would have easily dominated a typical eighty-square-foot office.[5] The mainframes commonly received input from punched cards like the ones Kurtz carried. A hole punched in the card at a particular location communicated a letter, number, or symbol to the computer, and each card featured several rows of punches. A computer operator loaded the cards into the computer to run the program. The computer communicated its results through more punched cards or magnetic tape or, most commonly, printouts.[6] In addition to being large, the mainframes were also very fast and very expensive. MIT's IBM 704 performed four thousand operations per second.[7] In 1962, GE priced one of its average mainframe computers, the GE-225, and its auxiliary equipment at nearly $240,000—close to $2 million in 2018 dollars.[8] Thus, any institution that had purchased or leased a mainframe aimed to keep it running as much as possible, to maximize its return on investment.

A carefully ordered set of punched cards often represented the culmination of the programming process. A mathematician like Kurtz first handwrote a program, either on scrap paper or in a special programming notebook. The notebook featured demarcated columns where the program author could write in commands and data that would be understood by the computer. The programmer could also make notes on what each step of the program was meant to accomplish. The columns were visual cues for converting handwritten notes to punched cards. In some cases, program authors punched their own cards using a keypunch machine. By 1958, Dartmouth had installed IBM keypunch equipment for its accounting operations, so Kurtz and his colleagues punched their own cards.[9] In larger programming operations, the program author submitted handwritten programming notebook pages to a keypunch operator who would then punch the cards. Kurtz would have spent hours working out a complex program. After he translated his program onto punched cards and ran it, additional hours or days would be needed to address any errors—to debug the program.[10]

Numerous errors crept into this programming process. A misplaced period—a simple dot—written into the code and punched in the card could dramatically alter the results of a program. A hole punched in the wrong location on a card could create an error. Indicating division instead of addition for a particular programming function could wreak havoc with a program. If Kurtz produced a computer program to perform a series of mathematical operations, and at some point the program told the computer to divide by zero (an operation not anticipated by Kurtz), that would have been an error. Typos, punch errors, misplaced punctuation—all of these confounded programmers, as did the challenges of communicating with the computer via complicated programming languages.

If Kurtz had been the sole user of the computer while he programmed, the very fast and very expensive computer would spend only seconds, maybe minutes, actually running his program. And if Kurtz had been the sole user, the minutes during which he loaded his punched cards and waited for the computer to print his results would have been minutes during which the computer's central processor was not active—costly minutes lost to inactivity. Thus, at the New England Computation Center at MIT and at other university computer centers during the latter 1950s and through the 1960s, computer managers focused on how to most effectively use the scarce and expensive resource of computer processing.[11]

As a solution, computer operators organized groups of individual programs to be run together, one group after the next, with as little computer downtime as possible, to maximize computer utilization. These groups of programs were known as batches, and this method of using mainframe computers became known as batch processing. Batch processing kept the computer humming, but it left programmers waiting hours or days for results. A GE consultant offered this description in 1963:

> If we follow a particular job [program] through this procedure, we find that the job is still waiting for its turn at all of these various manual input-output operations. It waits for key-punching, it waits for the batch to be collected to be put on the computer, it waits until the computer finishes processing all of the other jobs in the batch, it waits for the batch to be printed, it waits for someone to de-collate the combined output of the many jobs, and then it waits to be mailed or sent back to the man who had the problem run.[12]

Nonetheless, the various scientists, engineers, and business managers who relied on batch processing knew that a computer operating this way would still yield solutions faster than no computer at all. "So everyone puts up with this computer accessibility bottleneck," the consultant concluded, "and wishes it wasn't there."[13]

Batch processing frustrated Kurtz and his colleagues in the mathematics department at Dartmouth. Dartmouth College is located in Hanover, New Hampshire, not far from the Connecticut River, which forms the natural border between Vermont to the west and New Hampshire to the east. Appalachian Trail through-hikers trek through Hanover on their journey from Vermont's Green Mountains to New Hampshire's White Mountains. One observer described the "endless procession of mountain lakes" nestled in the region's lush green hills as a "year-round vacationland."[14] The Dartmouth community embraced the region's recreational possibilities through its student-established Outing Club. In 1956, just a year before MIT formally dedicated its Computation Center, the college opened its own ski area, the Dartmouth Skiway. Dartmouth students further cultivated their brand of masculine, rugged athleticism by according football a prominent place. In 1966, a correspondent for the journal *Nature* characterized "this isolated university town" as "best known for the success or at least the roughness of its football team."[15] Dartmouth is the northernmost member of the Ivy League, a conference formalized by the formation of the National Collegiate Athletic Association (NCAA) Division I in 1954. This group, which also includes Brown, Columbia, Cornell, Harvard, the University of Pennsylvania, Princeton, and Yale, has long been synonymous with elitism, socioeconomic privilege, and power. Dartmouth did not admit women until 1972, and during the 1960s, when the undergraduate population hovered around three thousand, students were overwhelmingly white and affluent.

John G. Kemeny was recruited to join the Dartmouth mathematics department in the mid-1950s when the dean of faculty, Donald Morrison, realized that nearly all its professors were approaching retirement. He recognized this as an opportunity to gather a new group of young and highly talented mathematicians as part of his plan to raise Dartmouth's academic reputation.[16] With the help of his colleague Allan Tucker at Princeton, where Kemeny earned his bachelor's degree in 1947 and his doctorate in mathematics in 1949,

Morrison was able to hire the bright young mathematician.[17] Kemeny was born in Budapest, Hungary, in 1926. With his parents, he fled the Nazi persecution of Jews by emigrating to the United States in 1940. As an undergraduate, he spent a year working in the theoretical division at Los Alamos during World War II, where he gained experience with computing.[18] At Princeton, Kemeny also worked as Albert Einstein's research assistant.[19] Kemeny joined the Dartmouth faculty for the 1953–1954 academic year but arranged for that first year to be a sabbatical. He assumed the chair of the mathematics department in 1955, after only a year of active teaching at the college.[20]

Kemeny exhibited broad interests, and he was fluent and comfortable conveying his sometimes unorthodox ideas. During his first decade at Dartmouth, he published three math textbooks and the monograph *A Philosopher Looks at Science*, dedicated to Einstein. He also published essays on math, education, libraries of the future, artificial intelligence, and computers in the *New York Times Magazine*, the *Atlantic Monthly*, and other popular periodicals.[21] Kemeny's writing demonstrated that he was not just a mathematician. He had been thinking a lot about the nature of education, students, and society. He asserted that "mathematics, education, and computers are three of the significant forces influencing our civilization."[22] He translated that conviction into action with his choices at Dartmouth during the 1960s.

Kurtz was a fellow Princetonian a few years younger than Kemeny. Kurtz's research focused on statistics, but he gained experience with computing at Princeton. He "would have gone into computing" if it had been an academic field at the time, and he carried that enthusiasm with him to Dartmouth.[23] Kemeny recruited Kurtz from Princeton to the small college in the New Hampshire mountains. In fact, the woodsy countryside appealed to Kurtz more than the college's math department.[24] Kurtz arrived on campus for the 1956–1957 school year, and Kemeny encouraged him to act as the liaison to the New England Computation Center at MIT the same year.[25] Access to the IBM at MIT enabled faculty members to experiment with computing via the tedium of batch processing, but Kemeny wanted more than that for his mathematics department.

Thomas Kurtz *(left)* and John Kemeny focused on users' needs in developing the Dartmouth Time-Sharing System, later known around New England as the Kiewit Network. Courtesy of Dartmouth College Library.

Kemeny obtained a computer for Dartmouth, in large part to attract mathematically talented undergraduates to his department.[26] In 1959, visitors to the college witnessed the construction of two connected buildings just north of the iconic Baker Library. Bradley Hall and its brother, Gerry Hall, rose from the ground as rectangular boxes covered in tiled panels of blue, white, and green, which earned the buildings the collective nickname of the "Shower Towers."[27] The math department would move to its new location in Bradley when the building was completed in 1961. Kemeny persuaded Morrison, now the provost, to allocate funds from the Bradley Hall budget for purchase of a small mainframe computer.[28] They obtained an LGP-30.[29] Although it was a "small" computer, the Librascope General Precision (LGP) machine measured about 20 cubic feet and weighed 800 pounds.[30]

The small computer, dedicated to Kemeny and Kurtz's purposes, reversed the logic of the large mainframe computer that MIT operated. Now, instead of working based on the idea of optimizing the machine's time, they would be able to optimize their time. Through their collaborative computing experiences, Kemeny and Kurtz became increasingly convinced that their Dartmouth students could—and should—learn how to work with computers. Undergraduates taught themselves the machine language necessary to program the LGP-30, and they wrote mathematical programs for it.[31] They joined an LGP-30 users' group and eagerly absorbed the ideas and suggestions shared in the group newsletter.[32]

The projects that Kemeny and Kurtz completed on the LGP-30 alongside their students convinced them that they needed to further expand computing options at the college. Kurtz worked with four students to program an algebraic language compiler for the LGP-30 from scratch, an impressive project from a student group.[33] Two of those students went a step further and created a faster, simplified version of that compiler for widespread student use.[34] Kemeny and Kurtz estimated that several hundred students learned and utilized this system, called SCALP, between 1962 and 1964.[35] During the 1961–1962 year, Kemeny collaborated with a first-year student, Sidney Marshall (class of 1965), to create a very simple programming language for the LGP-30. They named the project "Dartmouth

Oversimplified Programming Experiment," or DOPE.[36] Moreover, DOPE represented "an experiment designed to teach a large number of freshmen the rudiments of programming in a course of three one-hour lectures."[37] The experiment succeeded, which boded well for "the much larger-scale educational experiments planned for the future."[38]

A Guiding Principle for Dartmouth Computing: Users

Kemeny and Kurtz believed that computers would soon become a part of daily work life, possibly even a part of personal life. They explained, "Obviously, the coming generation must be prepared to understand computers and to make the most of them. Some meaningful contact with high speed computers should therefore be part of the liberal education of students."[39] This conviction, combined with their students' prowess on the LGP-30, propelled them to consider making computing accessible to many more students.

Even though their direct access to the LGP-30 enabled them to accomplish far more than they had at MIT, computer usage was still limited. The LGP-30 could run only one program at a time, with the associated problems of manual input for that program via punch card or magnetic tape. Moreover, smaller computers were slower computers, and even elementary problems could take a long time. An additional limiting factor involved compilers, by which the computer "translates or compiles" the program from something written in a "simple intuitive language" to "a form more palatable to the machine itself."[40] Kemeny and Kurtz knew that use of a compiler would make computing more accessible for their undergraduate students; however, small computers such as the LGP-30 could not offer compilers as "sophisticated and easy-to-use as those prepared for large computers."[41] They were also frustrated with FORTRAN, the common but complex programming language of the day. They knew their students exhibited a passion for computing, and they had personally experienced the productivity and satisfaction of programming. They had acquired the literacy to machine code a mainframe, and they had dabbled in the creation of programming languages with DOPE.

In either 1961 or 1962, Kurtz approached Kemeny with a three-pronged proposal, the ideas of which were deeply intertwined.[42] First, Kurtz wanted all of the students at Dartmouth to have access to computing. Second, Kurtz envisioned that access to computing would be free to students, a privilege comparable to their open stack library access. Finally, he believed the first two features could be accomplished through a new mode of computing called time-sharing.[43] Kurtz had learned of time-sharing from his former Dartmouth colleague John McCarthy. McCarthy, a pioneer in the field of artificial intelligence, had worked at Dartmouth during the 1950s and moved to MIT during the 1960s.[44] At MIT, McCarthy collaborated with Fernando Corbató's research group, working on implementing time-sharing.[45] Corbató and his research group presented their efforts at the Spring 1962 Joint Computer Conference.[46] The guiding concept for time-sharing was that multiple users could run programs on a mainframe computer at the same time.[47] Corbató and McCarthy envisioned multiple terminals connected to the mainframe, with a person at each terminal. The user could input and submit the program directly to the computer, then receive results shortly thereafter. If the results included error messages, the user could continue at the terminal and work to debug the program. Rather than taking days or even weeks to write and debug a program, time-sharing promised programming and debugging in minutes.

In developing the concept of time-sharing, McCarthy, Corbató, and their colleagues capitalized on the speed and efficiency of early 1960s mainframe computers. Mainframes processed simple programs in fractions of a second. Even if ten or twenty users submitted their programs to a time-sharing computer at the same time, they could all receive their results within a second or two, nearly instantaneously. A mainframe programmed for time-sharing accomplished this with a master, or supervisor, program that interrupted its own processing of user-specified programs to check whether other users had submitted requests. This master program effectively scanned for new requests a few times per second and responded accordingly. In their vision of time-sharing, Corbató and his group prioritized shorter programs over longer ones in terms of response time. If the computer tackled a particularly complex program, the response time was accordingly longer, perhaps several minutes. The response time increased not only because the computer required more time for computations, but also because the long program could be interrupted multiple times for shorter user requests. McCarthy suggested to Kurtz, "Why don't you guys do time sharing," then Kurtz shared the idea with Kemeny.[48] Kurtz researched the other institutions in the United States that were implementing time-sharing, and he communicated with Corbató about the MIT project.[49] Several other universities, including MIT, were working on time-sharing, but their projects focused on the university's scientists, mathematicians, and engineers as users. Although Corbató envisioned time-sharing as an opportunity for campus-wide use of computers, including by students, MIT time-sharing remained the domain of a self-selected group of scientists, engineers, and programmers throughout the 1960s.[50] Moreover, MIT's time-sharing system in 1962 enabled only three simultaneous active users.[51] When Kurtz brought the idea of time-sharing to Kemeny, the two developed a vision of time-sharing that was quite different from what had been achieved at MIT.

Kurtz also insisted that computing access should be available to all students at no cost. Kemeny shared Kurtz's enthusiasm for student access, which time-sharing would make possible. These paired ideas shaped the future and culture of Dartmouth for decades to come. When Kurtz likened computing access to library access, he reasoned that Dartmouth was responsible for buying books and maintaining a library to which students (and faculty) had borrowing privileges, simply by being at the university. Student access added little cost to an already substantial capital investment in learning and knowledge. Similarly, Kurtz contended, if the university invested in a computer to advance learning and knowledge, students should also have access to that computer.[52] The analogy between computing and library privileges went further. Kemeny summarized,

> The standard operation of a computing center is that of the closed-stack [library] method, where the users hand in a problem, have to wait a significant amount of time to have it serviced, and have answers handed back to them. They can communicate

> with the computing machine only through intermediaries. Our proposed solution amounts to an open stack operation of the computing center.[53]

Kemeny and Kurtz began to seek institutional and financial support for their endeavor, and to evaluate potential computer vendors.

The Dartmouth group considered computer providers during 1962–1963, and in the process of choosing GE, they developed a novel two-computer approach to time-sharing. They began researching providers before they had received formal financial support from either the college or the National Science Foundation (NSF). However, it seems likely that they received assurance of some financial backing during 1962 because they proceeded so confidently. They attested that, at the time, they "were very fortunate to receive full support" from President John Sloan Dickey, Dean Leonard Rieser, and Dean Myron Tribus of the Thayer School of Engineering at the college.[54] They investigated the offerings from Burroughs Corporation, Bendix Corporation, GE, and IBM.[55] As Kemeny and Kurtz considered GE's product line, they realized that GE offered two computers that would suit their time-sharing needs, the GE-225 in combination with the Datanet-30 communications processor.[56] The Datanet-30 could receive and manage incoming requests from multiple input devices (such as teletypewriters), and could then relay those requests along to the GE-225 mainframe computer for processing. After the GE-225 had finished running a particular user's program, the Datanet-30 could receive the results from the GE-225 and forward them to the appropriate user. This approach to time-sharing rested on the Datanet-30's programmable internal clock.[57]

McCarthy and Corbató at MIT had realized that the key to creating a time-sharing system was writing a clock or counter into the central processing unit of the computer; GE essentially offered a separate clock in the form of the Datanet-30.[58] During the mainframe era, once the computer began running a program, or a batch of programs, it continued running until it was finished with the entire program or the entire batch. Time-sharing required giving the central processing unit the ability to interrupt itself at some specified interval (such as 10 times per second) to scan for new incoming programs.

Corbató thought that could be accomplished without disturbing the program that was running at the time of interruption. Through the use of a clock or counter, a time-sharing system could place one program on hold while another received priority. In some cases, the computer bounced back and forth between programs, completing some of one, then another, so that both users received results at approximately the same time.

Thus, the name "time-sharing" resonated on two levels. The users were sharing the resources, and therefore the computing time, of one mainframe computer. The computer itself was modified with the addition of a programmed clock or counter, which regularly scanned for new user requests several times per second, and allocated—or shared—processing resources accordingly. Anthony Knapp, a Dartmouth undergraduate who worked with Kurtz on the GE negotiations, explained, "The ability of the system to interrupt itself at intervals would be critical for implementing time sharing; otherwise one user with a huge program could monopolize the GE-225."[59]

The Dartmouth group exhibited unorthodoxy in considering a two-computer solution for time-sharing. GE had designed and sold the Datanet-30 as a dedicated communications machine, with special-purpose software written by GE. Until the Dartmouth group approached GE about time-sharing, the company had not considered the Datanet-30 together with the GE-225 as a time-sharing solution.[60] Dartmouth's plan to use the Datanet-30 for time-sharing meant that Kemeny, Kurtz, and their students faced the added challenge of working with a communications machine that lacked a programming manual. Kemeny recalled, "At that time, many experts, at GE and elsewhere, tried to convince us that the route of the two-computer solution was wasteful and inefficient."[61] They pressed on with their vision.

Kurtz and Knapp traveled to GE's computing center in Phoenix to persuade GE to enter into a research partnership with Dartmouth. Kurtz and Kemeny hoped that GE would provide the computer equipment for free in exchange for Dartmouth faculty and students providing the programming and expertise to implement a successful GE time-sharing system.[62] To prepare for the visit, Knapp and Kurtz coauthored a document outlining how the GE-225 could be used together with the Datanet-30 for time-sharing.[63] Knapp remembered, "I threw something together in a few days, using a keen new plastic stencil for making balloons and rectangles and stuff that could be arranged into a professional-looking block diagram. The block diagram outlined steps in how a time-sharing program ought to work."[64] However, GE rejected the partnership, and its October 1962 proposal to Dartmouth viewed the college as a customer rather than a partner.[65] Nonetheless, GE offered the college "a very substantial educational discount" of 60 percent off the purchase of the GE-225 Computer System.[66] After considering and rejecting proposals from the other companies, including the industry leader, IBM, the Dartmouth group indicated their commitment to GE in April 1963.[67]

As Kemeny and Kurtz finalized their negotiations with GE, the pair also worked to persuade the college's administration to approve and fund their plans. In April 1963, Kemeny delivered a speech to the board of trustees in which he argued that computing would be a necessary skill for all of Dartmouth's graduates, regardless of their professional paths.[68] He underscored that their immediate access to computing resources was a worthy goal for the college. Kemeny's conviction on this point set him apart from many other scholars and computing enthusiasts at the time. Better-known computer visionaries such as J. C. R. Licklider and Douglas Engelbart envisioned futures of computers in the hands of individuals, but they pursued their ideas through small research labs. They worked to improve the technology itself, rather than figuring out a way to bring existing technology to as many people as possible.[69] In contrast, Kemeny displayed a passion for widespread computing born from witnessing the creativity and enthusiasm of a handful of engaged students.

Kemeny's near-decade of experience at the college shaped his speech for the trustees, just as the institutional circumstances of the college itself shaped the proposal. Kemeny was not just a mathematician. His work on the nature of education, students, and computers had led to articles such as "The Well-Rounded Man vs. the Egghead" in the *New York Times Magazine* and "Teaching the New Mathematics" in the *Atlantic Monthly*, and on computer simulations in *The*

Nation.[70] He was comfortable persuading others. Similarly, Dartmouth's status as a liberal arts college facilitated the acceptance of "open stack" computing. Kurtz recalled, "Achieving open access was aided by the virtual absence, in those days, of government-supported research at Dartmouth. We were thus freed from the major constraint found at other large institutions . . . to charge students and faculty for computer services lest income from government grants be jeopardized."[71] Kurtz and Kemeny turned the lack of government-funded research at their college into an asset. They could provide widespread computing to their students for free without worrying about extensive accounting. Finally, Kemeny and Kurtz had always enjoyed the support of the college's president and two prominent deans.[72]

In his presentation to the Dartmouth trustees, Kemeny advocated a computer center that would maximize the convenience of the user; that would be the guiding principle for the Dartmouth Time-Sharing System (DTSS). He explained that many campus computing centers, built around powerful mainframes and batch processing, were organized to maximize the potential of the computer, with little consideration of the potential of the human using the computer. In contrast, Kemeny and Kurtz sought to maximize human productivity in collaboration with the machine.[73] Kurtz eloquently summarized that they wanted time-sharing "so we would bring computing to the people, rather than bringing the people to the computing center."[74] With this speech, Kemeny publicly emphasized that computing at Dartmouth would be focused on users, an unusual stance for a college computer center at the time.

While Kemeny turned to the trustees, Kurtz sought funding from the NSF, another process that codified a key feature of the time-sharing system. One application to the NSF Mathematics Division requested $525,000 "to establish at Dartmouth College a single institution-wide Computation Center based on a large central computer with multiple input-output stations and operated under a Time-Sharing system."[75] A related application to the Course Content Improvement Section requested $275,000 "toward the development of materials and computer programs for the integration of computing ideas into freshman mathematics."[76] Thus, by April 1963,

Kemeny and Kurtz had also decided that they would integrate computing into the college curriculum via first-year mathematics courses. The NSF Mathematics Division application underscored the advantages of both a time-sharing system and an open access system. When the college formally executed a contract with GE during the fall of 1963, it had not yet received a response to its NSF proposals. Kurtz recalled that the trustees agreed to pay for a one-year rental of the equipment with a purchase option. When they received notification of an NSF award in January 1964, they converted the contract to a purchase agreement before GE delivered the computers.[77] The NSF funded these requests through one three-year grant for 1964–1967.[78]

When Undergraduates Built a Computing Network

Between their April 1963 requests for funding and the scheduled February 1964 delivery of the GE computers, Kemeny, Kurtz, and their students refined and developed plans for the Dartmouth system, and the idea of a new programming language emerged. Kurtz had easily convinced Kemeny about time-sharing and free access for all students, but Kemeny had a hard time selling Kurtz on the idea of writing a new programming language. Kurtz believed that they could make some version of the then-popular language FORTAN work for their time-sharing system and their students. He worried that a Dartmouth-specific language would not be useful for their students.[79] Kemeny's earlier experiment with DOPE convinced him that they could develop a more user-friendly language than FORTRAN.

Kemeny envisioned a language that was especially easy to learn and use. He wanted students to be able to interact with the computer by writing their own programs as quickly as possible. He understood the satisfaction of seeing a successful program run, and he knew that would draw students into computing. He convinced Kurtz, and together they planned for Beginners' All-purpose Symbolic Instruction Code, or BASIC (Chapter 3 is devoted to BASIC).[80] During the summer of 1963, Kemeny developed the compiler for BASIC, which translated the commands entered by a programmer into commands the GE computer understood.[81] BASIC and time-sharing were intertwined from the beginning at Dartmouth, and that relationship became crucial in shaping the entire computing endeavor.

Kurtz emphasized that creating a working system that was usable by students and faculty was far more important than tinkering with the theories and intricacies of high-powered computing. During the latter half of 1963, the DTSS acquired a number of defining characteristics before the actual computers were even delivered. Kurtz specified these in a memo intended to chronicle the system's development.[82] He prioritized a system that was "externally simple and easy to use for the casual programmer."[83] He also recognized that the perfect could be the enemy of the good: "In other words, it is strongly felt that we should not attempt to invent the ideal Time Sharing System, but to content ourselves with a simple but working approximation."[84] The open access requirement meant that Kemeny and Kurtz had to design a system to optimize the number of terminals connected to the computer while maintaining a reasonable response time for each user.

The choice to elevate human convenience and ease of use over computer efficiency influenced programming choices. Kurtz instructed his programmers, "In all cases where there is a choice between simplicity and efficiency, simplicity is chosen. Every effort will be made to design a system convenient for the user. But maximizing the time that the 235 is used is *not* one of the goals."[85] Kemeny and Kurtz demonstrated their commitment to users in their memo outlining the time-sharing system. Rather than starting with an explanation of the computer processes involved, they explained time-sharing from the user's point of view—how a user accessed the machine, the commands entered by the user, and how the user exited the system. Only after that description did they detail the computer processes connecting the teletypewriters, the Datanet-30, and the GE-225 computer.[86]

The prioritization of ease of use extended to BASIC. Kemeny worked to write a language that could be learned in layers: once a student had mastered a beginner's level, with which he could still write powerful programs, he could supplement his skills with additional commands and programming techniques, but nowhere along the way would his advanced techniques detract from ease of use for the beginners.[87] These values of widespread use, user convenience, and a privileged position for BASIC—the values of individualized, interactive computing—became embedded in the system.[88]

During the fall of 1963, Kemeny and Kurtz refined one more parameter of their project: once the time-sharing system was running, all first-year students enrolled in mathematics courses would learn how to program.[89] Nearly all of Dartmouth's incoming first-year students took math. Kemeny and Kurtz understood that a programming requirement was an extremely effective way to introduce hundreds of students to computing. Notably, their goal was not to make their students into computer programmers but to familiarize them with computing, to introduce them to the ways in which computing could help them with their coursework, challenge them, and entertain them.

From the outset, the math professors aimed not for computer scientists but for computing citizens. They sought to familiarize Dartmouth students, who were expected to become "executives or key policy makers in industry and government," with computing.[90] They planned that the time-sharing system would be productively used by thousands of undergraduates with diverse academic backgrounds and interests.[91] Now they just had to wait for their GE computers, program a time-sharing system from scratch, write a new programming language—and make sure everything worked. Kemeny recalled, "The [NSF] reviewers were right that we didn't know how difficult a job we were undertaking. If we had, we might never have tried."[92]

After receiving the NSF grant, Kemeny, Kurtz, and their student programmers, including John McGeachie and Michael Busch, still had much work to do before the arrival of the computers on campus. They received manuals for both the 225 / 235 and the Datanet-30 from GE and began to learn all about the computers on which they would implement time-sharing.[93] In fact, the 225 / 235 manual also included manuals for the peripheral equipment that Dartmouth would be using with the computer: a punched-card reader, a "mass random access data storage system" that used magnetic discs, a high-speed printer, and a magnetic tape subsystem.[94] The Dartmouth team needed to learn the intricacies of machine programming for the 225 / 235 and the Datanet-30. These included commands such as

ADD (performing addition of two numbers), SUB (performing subtraction), ADO (a positive one [+1] added to a particular number in memory), and STA (for "Store A," whereby the data labeled "A" are placed into memory at a particular location).[95] Even without the computers, the team could begin drafting their programs for time-sharing and BASIC on the GE programming tablets.[96]

GE delivered the two computers in March 1964, and the Dartmouth team redoubled their efforts.[97] Dartmouth housed the computers in the basement of College Hall, at the heart of campus. The college placed the teletype terminals on the first floor of College Hall. Student John McGeachie acquired responsibility for the GE-225, and student Michael Busch gained programming ownership for the Datanet-30.[98] They had to figure out how to make their respective machines accomplish their pieces of the time-sharing system and, more importantly, how to make the machines communicate with each other. They proudly took ownership of the responsibilities they had been accorded. Sometimes the machines were known as McGeachie and Busch, and sometimes the young men were known as 225 and 30.[99] Computing was personal. A decade later, Kemeny reminisced about how Busch and McGeachie reacted to the challenge of making the two GE machines communicate with each other. He explained, "The two of them took this terribly personally. It wasn't John's machine and Mike's machine; it was John and Mike who were not responding. And they would stand at opposite ends of the room and yell at each other at the top of their voices."[100] McGeachie sheepishly recounted an incident whereby he programmed the GE-225 to deliver an error message of either "Busch did it" or "I did it," depending on the circumstances. He was convinced that he would never see the error message "I did it," and felt "very embarrassed" when that message appeared the very next day.[101] Busch and McGeachie persevered, with the support and assistance of Kemeny, Kurtz, computing center supervisor William Zani, and a handful of other professors from the math department.[102]

Busch and McGeachie triumphed during May 1964.[103] They sat at separate teletype terminals. They each typed a short BASIC program, and they submitted their programs at the same time. A few

In 1964, Dartmouth College undergraduate Michael Busch *(seated)* worked with mathematics professor Thomas Kurtz *(standing)* to implement the time-sharing system on the GE computer installation in the basement of College Hall. Photo by Adrian N. Bouchard, courtesy of Dartmouth College Library.

moments later, they heard the distinctive slow-but-steady sound of two teletype terminals printing a result. Success! They considered this version of time-sharing "Phase 0."[104] The features of Phase 0 included a log-on sequence whereby the user typed HELLO to activate his session at the teletypewriter, and the command LIST whereby the user could see a complete listing of the program on which he was currently working.

Just two months later, the time-sharing team proudly announced, "In our never-ending struggle to bring you faster and more efficient service, we are pleased to introduce at this time PHASE ONE, a major revision of the Dartmouth Time-Sharing System."[105] Phase One featured a modified HELLO sequence and new commands such as RENAME, SCRATCH, NEW, and OLD, devised so that a user would not need to repeat the HELLO sequence more than once per session. Users were also informed that, "To end any possible confusion that may still exist, you may obtain a catalog of all the problems saved under your user number by typing ' CATALOG.'"[106] Thus, the early versions of DTSS included user numbers to differentiate individuals and to enable individuals to save their own work. The DTSS team created programming commands to simplify user operations. These actions manifested Kemeny and Kurtz's emphasis on user-friendliness.

The team toiled frenetically during the spring and summer of 1964 to prepare for the fall 1964 semester, when several hundred incoming first-year students would learn how to use BASIC.[107] Planning for the intertwined integration of time-sharing and BASIC into the curriculum had begun months earlier. Kurtz and Kemeny coordinated with other mathematics professors about how many lectures to dedicate to computing, what kinds of problems to assign, and how to evaluate the problems.[108] They decided on four one-hour lectures integrated into the various first-year mathematics courses. During the semester of the lectures, they required the student to solve three or four math problems using BASIC and the time-sharing system.[109]

The time-sharing team creatively applied computing to the problem of grading the programs of hundreds of students. After a student had written his program, tinkered with it and debugged it to his satisfaction, and thought it "completely checked out," he submitted it for grading by typing TEST on the teletypewriter.[110] This keyword activated a Dartmouth-produced program called TEACH, which evaluated the student's program against several test problems. The student received a notification if his program was not successful, and he continued with his programming efforts. If the student's program succeeded, he received a typed message "indicating that he has fulfilled the requirements of writing a program for that particular problem."[111] The student submitted this note to his course grader to be recorded. Kemeny proudly declared, "In this manner, the faculty time required in teaching computing is kept to an absolute minimum."[112]

Kemeny and Kurtz's creative approach to computing indoctrination displayed their zeal to demystify computing and make it practical for users. Embedding the computing requirement in essential first-year mathematics course ensured that more than 75 percent of each incoming class would learn computing. Guest speakers in each course could deliver the four lectures, rather than the course professor having to learn all of the material ahead of time. The TEACH program served two purposes: (1) students received immediate feedback on their performance—an incentive to keep trying, and (2) faculty members and course graders faced no additional work beyond recording the completion of the assignment. When the students arrived on campus in September 1964, Kemeny, Kurtz, and their students were ready.

Using the User-Focused System

By early October 1964, the DTSS featured connections to twenty-one teletype machines, and the first computing lecture had been delivered to students in several math courses.[113] In September, GE had delivered a GE-235 computer to replace the existing GE-225 computer. In fact, Kemeny and Kurtz had planned on the GE-235 for their system from nearly the beginning of their negotiations with GE, but when they learned it would not be available until autumn 1964, they agreed to accept the GE-225 in the interim.[114] Kurtz and his team installed the GE-235 "with a minimum of difficulty." Twenty of the teletypes were housed at Dartmouth, mainly in College Hall, and one was located at nearby Hanover High School.[115] Students enrolled in honors calculus and second-semester calculus (for those who had completed first-semester calculus in high school) received the computing lectures. Kurtz delivered some of the lectures himself, using a teletype that had been "mounted on a rollable platform for easy transport to Filene Auditorium," the large lecture hall in Bradley-Gerry (the Shower Towers).[116] Kurtz and his fellow computing lecturers employed a "portable thermofax machine" to transform the printouts from the teletype into overhead projector transparencies when they used the teletypes during their lectures. The first lecture introduced the concepts of time-sharing, computers, and programs. The lecturers described and demonstrated programs for calendars and baseball, and they concluded with a simple BASIC program that introduced seven key BASIC commands. The students received their manuals, and they were free to explore the time-sharing system.[117]

Jim Lawrie, a member of Dartmouth's class of 1968, embraced time-sharing and BASIC. After attending the computing lectures in his Math 5 course, he carried his BASIC manual with him over to College Hall and sat down in front of a teletypewriter. The teletype looked just like a typewriter, except all the letters were capitals.[118] Lawrie noticed that some special characters, like the addition and equal signs, could be typed using the SHIFT keys. He pressed the ORIG key to turn on the teletypewriter. He located the RETURN key, and he remembered from the lecture that he had to press the RETURN key to transmit his typing to the computer. A few keys to the left of RETURN, he observed the arrow pointing to the left (←), on top of the letter O key. This arrow (accessed by pressing SHIFT + the letter O) "erased" the last character typed, in case of a typing mistake. Lawrie soon realized that his typing mistakes remained visible on the teletypewriter paper, but those arrows told the computer to ignore the previously typed character. Thus, if Lawrie typed ABCWT←←DE, he would see ABCWTDE on the yellow paper, but the computer would read ABCDE. Browsing through the BASIC manual, he saw that he could use the ALT MODE key just to the left of the Q to delete an entire typed line if he made a lot of mistakes in that line.[119]

Lawrie typed HELLO on the teletype, then pressed RETURN, to begin his time-sharing session, and his exchange with the computer appeared as follows (underlined words indicate those printed by teletype):

	HELLO *RETURN*
USER NO.—	224488 *RETURN* [Lawrie entered his six-digit student ID number]
SYSTEM—	BASIC *RETURN* [To use the BASIC programming language]
NEW OR OLD—	NEW *RETURN* [NEW indicated a new program; OLD retrieved an existing program]
NEW PROBLEM NAME—	CONVRT [six letters and/or digits to name the program]
READY.[120]	

A Dartmouth student, intent on computing by teletype, connected to Kemeny and Kurtz's user-focused time-sharing network. Courtesy of the Computer History Museum.

He had just communicated with the computer, and he was amazed at how quickly the responses appeared on his teletype. He sat for a minute or two, flipped through his BASIC manual, and then started typing. He was hooked.

Conclusion

The laborious nature of batch processing, which was hugely inefficient for any one person involved, propelled Kurtz and his fellow Dartmouth mathematics professor Kemeny to revolutionize programming at the college between 1962 and 1965. With the help of motivated, engaged undergraduate students, Kemeny and Kurtz designed and implemented a computing network based on time-sharing. At its official launch during the fall quarter of 1964, twenty users could sit at their individual terminals and directly write and debug programs in a matter of minutes. Those twenty terminals were all connected to one mainframe computer. Kemeny and Kurtz's time-sharing system managed the programming requests of the twenty users in such a way that the majority of users experienced a response time from the computer of a few seconds. Time-sharing represented a dramatic change from the use of mainframes: the users interacted directly with the computer via their terminals, and they submitted their programs while using the terminals, rather than having to hand punched cards to an operator and wait hours for batch processing results. Time-sharing provided a much more personal experience of computing, connecting the individual directly with the terminal, and the terminal with the computer. Kurtz and Kemeny elevated user convenience in the design and implementation of their time-sharing system, and they repeatedly affirmed their commitment in their plans for and execution of the system. In short, Kemeny and Kurtz's focus on the user inspired Dartmouth's networked computing and BASIC. In turn, BASIC and time-sharing fomented personal computing at Dartmouth and beyond.

2

Making a Macho Computing Culture

During the 1974 Pioneer Day Session at the National Computer Conference, the students who had taught the computer, and created the Dartmouth network in the process, gathered to reminisce about the previous decade of Dartmouth computing. Sidney Marshall recalled the charged atmosphere in the public teletype room before football games "when everybody was up there trying to impress their dates with the computer." The Computation Center student workers used their powers to remotely command a terminal to play jokes on their peers. Marshall reminisced,

> You'd take control of the computer so they'd be talking to you. You got some very interesting conversations. . . . One of the questions they asked was "What's the score of the football game going to be?" I typed something very fast. 14 to 7. And I was right! And there was a big argument after the game up in that room. "The computer can predict things!" "No, it can't!" I never did find out who it was. I just heard reports of it.[1]

During the 1960s, Dartmouth students were all men, and, for many, their social lives revolved around football and fraternities. When computing entered that equation, the result was a decidedly macho computing culture. In fact, computing and football were intertwined in campus culture during the 1960s. The uses of Dartmouth computing, such as showing off to a date, also underscored the significance of normative man-woman pairings that would ultimately culminate in the Cold War nuclear family. Although women worked at the Computation Center, their public representation emphasized their roles as mothers and wives, rather than as valuable contributors to a growing and thriving computing network. Finally, although Kemeny and Kurtz believed that computing offered a tremendous opportunity for all their students, the ways in which it was deployed on campus offered spaces for students to create novel associations between computing, masculinity, and status, like when Marshall and his Computation Center buddies played practical jokes on their unsuspecting peers.

Dartmouth's dedication to accessible computing for all its users set its network apart from similar computing networks during the 1960s. Yet those users, including Tom Kurtz and Greg Dobbs, were almost always white men. Dartmouth's network cultivated computing citizens; however, that citizenship mirrored the college's demographics: predominantly male, white, and affluent. Although Kemeny and Kurtz intended computing as an equalizer for their students, attention to gender and sexuality shows how computing hierarchies emerged. The idea of the "user" at Dartmouth demonstrates how the term itself masked layers of gendered expectations. Moreover, the intersection of Dartmouth's student body and the campus's nascent network yielded a computing culture of masculinity, whiteness, and heteronormativity.

This chapter analyzes the nature of computing citizenship at Dartmouth during its first decade. It begins with the Kiewit Center, the pride and hub of Dartmouth's personal and social computing. The next section examines the processes by which computing at the college became masculine, including a close reading of the college's intertwined football and computing cultures, the use of computing to reinforce traditional Cold War gender roles, and the public messaging about the women who worked at Kiewit. The final section teases out the whiteness of Dartmouth computing, a characteristic that was submerged below Kiewit's self-identification as the hub of multiple, diverse networks.

Kiewit: The Hub of the Dartmouth Network

Kemeny and Kurtz banked on the success of Dartmouth computing. During the autumn of 1964, when members of the class of 1968 first received their introductions to networked computing, teletypes, and programming in BASIC (Beginners' All-purpose Symbolic Instruction Code), the college planned for the construction of a new building to house its burgeoning computation center. Alumnus Peter Kiewit, who became a construction magnate in the Midwest after his tenure in Hanover, contributed a substantial capital gift of $500,000—and his name—to the Kiewit Computation Center. Kiewit had been admitted as a member of the class of 1922; however, he remained at Dartmouth for only a year before returning to his father's construction business. Kiewit's contracting flourished during the Cold War; his business built Thule Air Force Base in Greenland as well as numerous Titan and Minuteman missile facilities around the United States. Kiewit finally earned his degree from the college, an honorary doctorate, in the spring of 1964—while plans for the computation center expansion were well under way—with a citation describing him as "builder and tunneler to the world."[2] Kiewit (pronounced kee'-wit), as the Computation Center became known, was formally dedicated during the conference "The Future Impact of Computers," held on December 2 and 3, 1966. Kiewit's opening merited two articles in the *New York Times*, as well as a front-page spread in the student newspaper, *The Dartmouth*, and seven pages in the college alumni magazine.[3] Kiewit enjoyed a central location behind Baker Library, which anchored the north side of the grassy green quad at the heart of the campus. Kiewit also sat at the top of Webster Avenue, better known on campus as "Fraternity Row."

During the 1960s, social life at Dartmouth revolved around its fraternities, which meant that the new Computation Center was situated in a place of prominence for socializing, entertaining dates visiting from the Seven Sisters Colleges, or popping in on the way back from a football game. When Kiewit opened, Kemeny confidently declared, "I'm quite certain that this center will rank with Baker Library and Hopkins Center [for the Arts, which foreshadowed the design of New York City's Lincoln Center] as one of the three

Students computing and collaborating at a time-sharing terminal cluster in the Kiewit Computation Center at Dartmouth College. The center lent its name to the Kiewit Network, another name for the Dartmouth Time-Sharing System (DTSS) across New England. Courtesy of Dartmouth College Library.

facilities having greatest impact on the entire campus, and that it will have a continuing impact throughout the indefinite future."[4] The modern and low-slung white concrete Kiewit Center featured climate-controlled space for its computing machinery ("enough air-conditioning equipment to cool more than thirty average-sized houses"), and offices for its staff, but its social heart was the large public teletype room on the south side of the building, where light flooded in from the roof-level windows.[5]

The descriptions of the new Kiewit Computation Center focused not on the building itself but on the defining characteristics of Dartmouth computing that had emerged in just two years: pervasiveness, accessibility, and creativity. "Dartmouth College is a campus gone crazy for computers," declared the lede of the first *New York Times* article about Kiewit.[6] The correspondent described the scene: one student debugged a program he had written to test a psychology hypothesis, another ran a program named "Xmas" to print out his Christmas cards, and a third young man called on a template program to print this letter to send home: "Dear Mom, I'm so busy studying for finals that I don't have time to write myself. . . . Send money."[7] Dartmouth students were enthusiastic and eager to work interactively with a technology that was, for much of the American population, as remote as the moon.

Moreover, as the *New York Times* reporter noted, all students received free computing time. It is worth underscoring here that the students (or whoever paid their tuition) had paid for the privilege of attending the college; however, the Dartmouth computing network contrasted with those of most other universities, at which users had to pay for their computer time, on top of tuition. The Kiewit Computation Center did charge faculty and other users for their computing time; however, the college also maintained its commitment to computing access with a special research fund to cover computing charges for those faculty who lacked the grants or other financial resources to pay for computing themselves.[8]

The accessibility, along with the ease of BASIC, engendered tremendous student creativity. In addition to crafting Christmas cards and form letters to parents, students created computer art; beginning in 1969, Kiewit held an annual computer art contest.[9] Still others

The public teletype room at Kiewit Computation Center, Dartmouth College. Community members gathered to compute, collaborate, create, and communicate in this popular space. Courtesy of Dartmouth College Library.

logged onto the network to play one of the many games available, including backgammon (BACKGAMN), simulated slot machines (BANDIT: "The user places his bet and 'pulls the handle' by hitting RETURN. The game is over when the user's balance reaches 0"), basketball (BASKETBL), bridge, checkers, chess, generic FOOTBALL (which was differentiated from "Dartmouth Championship" FTBALL, and the even "more elaborate" GRIDIRON), hangman, poker, roulette, battleship, simulated slalom skiing, tic-tac-toe, and multifarious others.[10] Kemeny, Kurtz, and the students who programmed the Dartmouth network had spread the gospel of computing to their people, and those computing citizens eagerly communed with the teletypes at the temple of Kiewit.

How Computing Went from Male to Masculine

Kurtz, the director of the Computation Center, used the language of "citizenship" to describe the relationship between a user and the user's computing network. A good computing citizen respected the college's ongoing computer memory limitations. For most of its first decade, the Dartmouth network perpetually struggled with its popularity, which manifested in the form of a lack of sufficient storage for user programs. The Computation Center's near-monthly *Kiewit Comments* newsletter frequently provided updates on the state of computing storage, and the Kiewit staff requested, reminded, and cajoled users to actively "un-save" unnecessary programs to free up precious memory for other users. Kurtz declared, "Users can exercise good computer citizenship by UNSAVEing [*sic*] all programs no longer needed."[11] Not that the users had much of a choice: by 1968, the purge period had decreased from twenty-one to fifteen days because of the "extremely critical (as usual) storage problem"; any program that had not been used in the previous fifteen days was deleted. Users were encouraged to preserve their programs on paper tape or punched cards to free up disk space.[12]

Kurtz considered this kind of attentiveness to the network—its machinery, fellow users, and shared programs—the hallmark of good computing citizenship. Promptly reporting any machinery or connectivity problems demonstrated good citizenship because it provided Kiewit staff the opportunity to fix the problem before it affected simultaneous users at other locations or subsequent users at the same location. Prank-calling teletypes showed quite the opposite. "Teletype offenders" were castigated for their "breach of user etiquette" and causing annoyance, and the newsletter offered detailed step-by-step instructions "to print out the telephone number of the offending teletype."[13] Students showed good computing citizenship when they shared terminals: one could revise his program while the other entered and ran his program.[14] The Kiewit staff also exhorted experienced users to contribute their programs to the DARTCAT program library for the benefit of computing novices: "We need the support of all users. . . . May we encourage you to submit these 'goodies.'"[15] Indeed, the staff's dedication to soliciting programs as a shared and maintained software resource (with over three hundred programs by 1970) fostered widespread use, acceptance, enjoyment, and sociability on the network.[16]

Kurtz conceived of computing citizenship at the college network level, but Kemeny—who became president of the college in 1970—looked beyond the Hanover hills, and he believed that computing was crucial to good American citizenship. Kemeny emphasized that computing was essential for a well-rounded liberal arts education, but, more importantly, that the time-sharing network was a worthwhile investment because Dartmouth men were destined to be the future leaders of the United States.[17] Herein lay several subtle and intertwined assumptions. First, Kemeny believed (as did others during the 1960s) that computing would become increasingly important and increasingly quotidian. He also thought that any business, government, military, or scientific leader would have to be computer savvy. Kemeny's conviction that computing experience was requisite for full American citizenship—for contributing to the commons, the community, and the democracy—was more unusual. And Kemeny's assumption of individual and college-wide responsibility to encourage personal computing because one's students were leadership bound was more unusual still. These convictions also glossed over the fact that Dartmouth's future leaders were a homogeneous bunch of white men: no women, few minorities. Here was a world of personal and social computing at odds with the social justice movements of the long 1960s.

Kemeny and Kurtz elaborated a very particular examination for computing citizenship: writing programs in BASIC. The math professors required that any student enrolled in a mathematics course in his first year (roughly 75 percent of the students) had to produce several math-related programs in BASIC on the time-sharing network to pass the course. Perhaps this now seems like an obvious way to introduce most students to computing; however, one can imagine other possibilities: introducing computing in a required English course, or requiring all students to complete a computing project of choice sometime during their first year. Although Kemeny associated computing with leadership, its introduction in math courses de facto associated computing with mathematics. Moreover, the math course requirement excluded a significant minority of students from a communal learning setting in which to test BASIC and computing. Some students continued to use BASIC throughout their college years, regardless of major course of study, and some simply played games like FOOTBALL on the time-sharing network, but they had been united as users: male college-age students, predominantly white and well-to-do, who could create programs in BASIC.

Despite the fact that Dartmouth men learned BASIC programming in their math courses, Kemeny, Kurtz, and the students themselves cultivated a decidedly noncerebral breed of masculine computing centered on games. During the 1960s, Dartmouth's membership in the Ivy League revolved around its football team, which won the league championship seven times during 1962–1971. That was a tremendous source of pride for the remote northern college and its students. When the special correspondent from *Nature* reported on Dartmouth computing, he mentioned the "roughness of its football team."[18] Playing football demanded physicality and strength; it was a bodily contrast to the mental act of planning a mathematical algorithm and programming a computer. Watching football games generated camaraderie and school spirit, with abundant displays of Dartmouth green, pennants, and banners. When Reverend Eleazar Wheelock founded the college in 1769, he decreed its mission to encompass the education of Native Americans; and Mohegan minister Samson Occom helped raise funds for the institution. Two hundred years later, the college used "the Indian" as its unofficial athletic mascot. Students cheered "Wah-hoo-wah" during football games, and a caricature of a Native American brave's profile—with facial paint, earring, and feathers—decorated the masthead of the student newspaper *The Dartmouth*.

Considering that culture of rough football and racist Native American appropriation, the naming of the student ALGOL compiler (created under the tutelage of Kemeny and Kurtz) with the acronym SCALP in 1962 could be read as the first step in the production of Dartmouth's macho computing culture. A group of white men appropriated the "Indian scalping" stereotype—based on the racialized attribution of ferocity, bravery, and savagery to Native Americans—from football for computing, signifying their attempts to invoke the physicality, spirit, and masculinity of football for their brainy work.

The Kiewit Center brought that rough-and-tumble masculine bonding into the teletype room with at least three versions of computer football games (FTBALL, FOOTBALL, and GRIDIRON), and with other computing games of sport and war; in fact, Dartmouth distinguished itself from most other universities by actively encouraging student gaming and recreation on the network. Dartmouth's digital football received pride of place in numerous publications about the college's network, and it was frequently referenced in the Computation Center's newsletter, the *Kiewit Comments*.[19] In fact, the cover of *Kiewit Comments* 2.10 was computer art that depicted Snoopy kicking a football. Kemeny had written FTBALL to commemorate Dartmouth's 1965 football win over his alma mater Princeton, in which the undefeated Dartmouth secured the Ivy League Championship.[20] When the students who built the Dartmouth network gathered with Kemeny, Kurtz, and a handful of other Kiewit staffers in 1974 for their Pioneer Day session, the employee Nancy Broadhead recalled that Kemeny frantically called her late one night to report "in absolute panic" that FTBALL wasn't working. Kemeny, who by 1974 was president of the college, responded, "We were probably trying to recruit a new football coach and that seemed terribly important."[21] The exchange revealed that Kemeny valued FTBALL so highly—and symbolically—that he was willing to call an employee late at night to report its outage.

Broadhead's recollection of this vignette to a national audience subtly highlighted the gender and power dynamics of computing at Dartmouth. A distinguished, tenured male math professor had called a female Computation Center employee at home after hours to complain that a game was not available. Broadhead still worked at Kiewit in 1974. She described her role there as "part-time operator/consultant and probably more appropriately also housemother."[22] Although Broadhead performed significant computing duties at Kiewit, her self-identification as a "housemother" invoked a role of femininity, maternal nurture, and care for the male staffers and students associated with the Computation Center. This connotation of chaperone and even housekeeper could well have been an adaptive strategy for Broadhead to highlight a traditionally feminine gender role and to soften her expertise as rule enforcer and adviser in a developing masculine computing culture.

Dartmouth students created their own connections between the social spaces of football games and computing; those connections reinforced contained Cold War gender roles of heteronormative pursuing men and wooed women. Many Dartmouth men recalled that they often brought dates to College Hall or Kiewit before or after football games to demonstrate their computing prowess. Football games provided the opportunity for women from all-women's colleges to travel to Hanover to socialize with the Dartmouth men. It is worth emphasizing that it was the normative man-woman pairings that were also highlighted, never the possibility of the range of queer pairings beyond those that would yield a nuclear family.[23]

Dartmouth users employed computing in very personal and creative ways to reinforce existing gender roles. Many members of the class of 1968—the first group required to learn BASIC—recalled some sort of courtship in connection with computing. Francis Marzoni used the time-sharing system to create a huge printout proclaiming, "HEY GIRL I MISS YOU" for his girlfriend who was attending college in New York. He rolled it up in a poster tube, mailed it off, and has since been married to her for over forty-four years.[24] Another planned to woo his Winter Carnival date by composing a romantic text for her and "making this BASIC program hold it in memory for the proper moment when [she] would see this printout and be overwhelmed by [my] computer prowess."[25] Yet another recalled flirting with his now-wife of forty-five years over the time-sharing system while he was at Dartmouth and she was a student at Mount Holyoke.[26]

Playing games like FOOTBALL and SALVO42 made computing both personal and social for Dartmouth students. They sat in the public teletype room playing individual games, but with the companionship of their computer center buddies; this sociability became more explicit and pronounced when the Kiewit staff added a multiplayer feature to FOOTBALL and other games. The November 1969 *Kiewit Comments* announced the possibility of connecting multiple terminals to one program, including FOOTBALL.[27] Although the newsletter announced this development with characteristic understatement, I must underscore the transformation: Dartmouth students could now sit at individual teletypes in Kiewit and around campus (or, as Chapter 3 describes, diverse secondary school and college students could sit at teletypes located around New England), and they could socialize, interact, and play together with these multiplayer games on the Dartmouth network. The September 1970 *Kiewit Comments* reminded new and returning students of this possibility, adding that blackjack (BLKJK) and POKER were additional multiplayer games.

SALVO42, like the various football games, epitomizes the machismo of Dartmouth computing. It was a multiterminal simulation of a naval battle, in which the objective was to sink the ships of one's opponent(s). The word "salvo" connotes not just the simultaneous discharge of guns or other weapons during battle; it can also refer to any vigorous or aggressive series of acts. Similarly, "42" invoked 1942 and the numerous World War II naval battles in which the United States engaged in the year after the bombing of Pearl Harbor. These themes of war and violence in computing games have been extensively analyzed for the ways in which they shape and reinforce gender roles.[28]

Another example of Dartmouth's social computing was the popularity of the SIGN and BANNER programs, which also merited frequent reports in *Kiewit Comments* and were deployed creatively to reinforce heteronormative Cold War gender roles. Initially, the SIGN program printed banners in enlarged letters; messages "up to 80 characters long" were "printed in letters 60 spaces high and 12 lines wide"—useful for making signs for football games, to announce fraternity parties, or to publicize other campus activities and events.[29] Less than a year later, Kiewit staff reported, "The popular program SIGN*** has been replaced by a better version called BANNER*** which offers a choice of two letter styles and three printing formats."[30] A student using SIGN or BANNER created strong tangible and visible connections between computing and college life. Celebrating birthdays, supporting friends in a ski meet, mocking one's football opponents, publicizing a fraternity gathering, welcoming the women who traveled to campus to celebrate Winter Carnival, or wooing a girlfriend: all these activities made computing quotidian.

During the 1960s, women were not undergraduates at Dartmouth but they were present as staff at the Computation Center; the labor that they performed, and the ways in which they were publicly described via *Kiewit Comments* reveal characteristics of the broader computing industry, as well as the evolution of the college's particularly masculine and heteronormative computing culture. The women who worked at Kiewit were employees of the burgeoning American computing industry. We may think that women were scarce in computing during the 1960s, but historical research has consistently demonstrated otherwise. Moreover, historians and other scholars have criticized the processes by which American society collectively erases the labor performed by women. In fact, the fields of computation, mathematics, and information processing had long employed women, from the women who worked as computers at the Harvard College Observatory and the women who worked as telegraphers during the nineteenth century to the women who computed the Math Tables Project during the Great Depression, programmed the ENIAC during World War II, and calculated the astronauts' trajectories for the moon missions. Historians have also emphasized that women were often paid less for performing the same work as men, and that their jobs were categorized as lower status, less important pink-collar jobs only because they were performed by women. In American computing during the 1960s and 1970s, women were gradually pushed out as the field professionalized, a process intertwined with the creation of a particularly masculine computing identity.[31]

Teletype usage at Dartmouth during the 1960s offers a microcosm of the ongoing erasure of women's work in the history of information processing. By 1965, the teletypewriter—the terminal on which Dartmouth students wrote their BASIC programs—had a fifty-year history in American communication. The teletype had been introduced on the telegraph system around 1910 as a way to input messages via typewriter rather than Morse code. In fact, Western Union—the telegraphy giant—opened schools across the United States to train women on how to use the teletypes.[32] Teletypewriter work quickly became a pink-collar field—low-status clerical work performed by women. Despite the fact that the teletype was firmly fixed in the realm of women's work by 1965, it shed those gendered connotations at Dartmouth. The Dartmouth men embraced the teletypes and time-sharing as their own, focusing instead on the difficulty of pressing the keys, the noise of multiple teletypes in simultaneous use, and the modern architecture of the new computing center.[33] Indeed, the use of teletypes with the time-sharing system at Dartmouth (and the subsequent use of teletypes with minicomputers and personal computers during the 1960s through the 1980s) seems to have rendered invisible that earlier—and women-focused—history.

Yet, the women of Kiewit epitomized the range of possibilities for women in professional computing during the 1960s: application programmers, operators, technical librarians, computing program coordinators, and secretaries. Janet Price joined as an applications programmer in 1968. Price not only served as an expert on the FORTRAN programming language but also lectured on it; she also developed programs for the Dartmouth network for college faculty. Price started programming in 1960, while she was earning a bachelor's degree in mathematics from the University of California at Los Angeles. At Dartmouth, she completed a doctorate in psychology in 1971 with a dissertation on mathematical models in cognitive psychology.[34] Although Price's progression from a mathematics degree to employment as a computing expert to a doctorate may seem unusual for the 1960s, women often had important roles in computing—from keypunch operators to programmers—during that decade. From the opening of the Kiewit Computation Center through the 1980s, Nancy Broadhead wore many Kiewit hats, from operator to consultant to manager of user services, sharing her expertise with the publication of two articles.[35] Jann Dalton worked as the Kiewit librarian, and her responsibilities included editing the *Kiewit Comments*.[36] Ruth Bogart joined as a social sciences programmer, supporting faculty and research projects in those fields with her computing expertise.[37] Diane Hills and Diane Mather joined the staff during the summer of 1969, jointly responsible for "applications programming, the DARTCAT library and other user services."[38] Prior to her Dartmouth employment, Hills had graduated from MIT and worked as a programmer at the Smithsonian Astrophysical Laboratory in Cambridge, Massachusetts. Mather had worked for two years in the computer center at the State University of New York at Buffalo. Mather revised the 1970 version of the BASIC manual for the Dartmouth network, and she coauthored the 1973 edition of the

DTSS program library.[39] This range of jobs represented a sample of the diversity of employment available to women in computing during the 1960s.

What is notable in the case of Dartmouth employment is that the *Kiewit Comments* differentiated the married status of "Mrs. Janet Price" from another applications programmer who joined at the same time, "Alicia DeNood." DeNood had earned a bachelor's degree in statistics in 1967 at Radcliffe College, and she had worked as a programmer for the Goddard Space Flight Center, the type of labor recently highlighted by the book and film *Hidden Figures.* The distinction between "Mrs. Janet Price" and "Alicia DeNood" emphasized Price's married status. I call attention to this because it was not an isolated incident; it was one of many examples of how the *Comments* differentiated between married women and those who were either not married or preferred to not be identified as "Miss" or "Mrs."[40] The different signifiers for unmarried versus married women, compared with the universal "mister" for men, has long served a social system in which girls and women are foremost identified by their marital status.

The employment arrivals and departures of Kiewit women were often discussed in terms of their husbands or children, whereas the wives or children of the Kiewit men were rarely introduced. When editor Lois Woodard left Kiewit, the newsletter announced, "Lois and her husband, Mike, leave Dartmouth on June 17. . . . Mike will begin training in General Electric's Marketing Management Program."[41] When "Mrs. Susie Merrow" joined the staff, the newsletter added, "*Susie* and her husband, *Ed,* who is a senior at Dartmouth majoring in government, make their home here in Hanover."[42] Similarly, when *Comments* editor Jann Dalton birthed a daughter, the newsletter reported, "Congratulations to Joel and Jann Dalton on the birth of their 6 lbs. 8 oz. daughter Stephanie. . . . We hope to have our Editor back on hand for the next issue."[43] One might argue that this type of reporting contributed to a friendly, familiar atmosphere among the Kiewit staff; however, it also reinforced the gender norms and gender roles of the Cold War era. Women were elevated as wives and mothers above their professional computing contributions.

A Dartmouth student works on the upgraded GE 635 time-sharing system at the Kiewit Computation Center. Undergraduates implemented time-sharing on an earlier GE system, and they continued to play critical roles in supporting the college's popular network. Courtesy of Dartmouth College Library.

Meanwhile, the distinctive practice of students teaching the computer continued with the ongoing employment of Dartmouth students at Kiewit. Students interested in becoming systems programmers completed a monthlong "apprenticeship," after which they became "full-time student systems programmers."[44] In 1972, the Kiewit leadership proudly reported about their expansive New England network, "The bulk of the programming effort has been undertaken by Dartmouth undergraduates under the supervision of faculty members. . . . These students have worked part-time at the computer center during the academic year and full-time during the summer recess. This programming activity has been entirely extracurricular; the students have carried a normal undergraduate course load at all times."[45]

Because Dartmouth prided itself on its student system programmers, and because all of those students were men in the formative years between 1962 and 1972, Kiewit was dominated by young men, many of whom enjoyed the power associated with their employment and status. Although Kemeny and Kurtz may have envisioned computing as an equalizer among their students, the students perceived it differently. The students who had programmed the time-sharing system, and who continued to develop and maintain the mainframe computers, created a space and hierarchy for themselves based on their familiarity with the system behind the scenes. They mocked the students who thought of the teletypes themselves as the "computers." They played practical jokes by which they would randomly substitute strings of meaningless text into the output of someone's laboriously written program. They delighted in the arcane details of their programming expertise.[46] These Computation Center student employees had cultivated status—and created a particular form of masculinity—for themselves by understanding the obscure machine language required to communicate with the mainframe computers, by exerting power over their peers, and by flaunting their expertise compared with that of older computing professionals, such as the men employed by GE or IBM.[47]

Even after the college admitted women as undergraduates, very few (if any) sought employment at Kiewit; the first, formative decade of the Dartmouth network created a masculine computing culture for users and experts alike. Kemeny had advocated for and overseen Dartmouth's transition to undergraduate coeducation, beginning in 1972 (the class of 1976 was the first to include women). He hoped to open computing citizenship to these women. Yet, all thirty-six of the Kiewit student programming assistants employed during 1973 through 1976 had masculine names.[48] The 1976 report on computing at the college featured an abundance of photographs, in which for every seven men pictured, only one woman was pictured.[49] In other words, of the people visually representative of Dartmouth computing, more than 85 percent were men. That year, the student population was still overwhelmingly male, at 73 percent, so perhaps it is not surprising that Kiewit remained a masculine stronghold. Yet, as I argue in Chapters 3 and 4, Dartmouth computing influenced other spaces and places of computing citizenship in the 1960s and 1970s through its own network, through the national recommendations made via the President's Scientific Advisory Committee on which Kurtz and Kemeny served, and, above all, through BASIC. Indeed, Kiewit claimed that based on the "Dartmouth-like" time-sharing systems marketed by then-popular manufacturers Hewlett-Packard and Digital Equipment Corporation, "it seems safe to conclude that perhaps millions of students in the United States . . . have learned computing Dartmouth style."[50] And "computing Dartmouth style" had already become a decidedly masculine endeavor.

Finding Whiteness beyond the Black and White

The lack of racial diversity in Dartmouth computing was revealed in the silences about it, the assumptions of homogeneity. And the handful of examples of how and when race was invoked demonstrates the social construction of whiteness as the normative college computing culture. When the trustees were considering candidates for the next president of Dartmouth in 1969, one of them asked the journalist William J. Miller to visit the campus for a few days and record "his honest, candid opinion of . . . its flaws, its strengths, its opportunities." This manuscript, "A Visitor Looks at Dartmouth," provided an unusual perspective on the college; a number of faculty members and administrators—including the president, the provost, the dean, and Kemeny (who became the next president)—offered their observations on its strengths and weaknesses.[51] In his conversation with Miller, President John Sloan Dickey piled praise on the William Jewett Tucker Foundation, which had been established in 1951 "to give students a sense of conscience and commitment."[52]

In this context, Dickey mentioned the "A Better Chance" (ABC) program, explaining that it originated "when some of the prep schools six years or so ago wanted to offer more scholarships for disadvantaged youngsters, but had great trouble in finding enough who were able to do the work or keep up with it."[53] On the surface, this sounded like a scholarship program for students who could not afford prep school or Dartmouth tuition. But "disadvantaged" referred to low-income students, most of whom were African American.

Indeed, the civil rights movement had propelled the formation of the ABC program in 1963 to create a pipeline for minority students to succeed in preparatory schools. The following summer, about fifty students, nearly all of whom were black, attended an intensive summer school session at Dartmouth. Upon successful completion, those students were admitted, tuition paid, to elite prep schools including Phillips Andover, Phillips Exeter, Choate Rosemary Hall, Miss Porter's School, and many others. After graduation, many of the ABC students continued to prestigious colleges and universities including Dartmouth, Harvard, Smith, Stanford, Wellesley, Wesleyan, and Yale.[54] Dickey expressed pride in the ABC program; he estimated that (in 1969) about 350 ABC students had matriculated to colleges including his own, Amherst, Williams, and Mount Holyoke in Massachusetts, and Carleton in Minnesota. About a dozen ABC students entered Dartmouth in 1968, and about two dozen in 1969.[55] The Black Alumni at Dartmouth Association estimated that, altogether, about ninety black students matriculated in 1969, and enrollment remained about that level during the 1970s.[56] This marked a substantial shift in student demographics; the college had only a handful of African American students during most of the 1960s, and suddenly the numbers swelled. Nine percent of the class of 1976 was African American, coming close to reflecting the 11 percent of African Americans in the U.S. population in 1970.[57] As president, Kemeny continued to diversify the student population by pushing for the admission of women and by recruiting Native American students.

Despite this influx of minority students at the end of the 1960s, Kiewit reporting that highlighted the ABC program ultimately called attention to the differences between ABC students and the white student majority. This reinforced the tacit co-construction of Dartmouth computing with whiteness by emphasizing the ABC students as different. The September 1968 *Kiewit Comments*, which welcomed the class of 1972 and provided them with a primer on Dartmouth computing, reported on "Summer BASIC Instruction for 'ABC' Students." The newsletter did not explain what the ABC program was, but it noted that the 121 students who participated at Hanover High School "became comfortable with computer based instruction and recognized some of the possibilities and limitations of this marvelous facility."[58]

Although President Dickey's comment about "a group of students who have set up a special ABC tutoring program in Jersey City, to get to the heart of the ghetto" was not a public comment, an extensive *Kiewit Comments* report about students working in Jersey City was.[59] A mixture of Dartmouth undergraduate and graduate engineering students lived in Jersey City, working as tutors, laboratory assistants, and teachers' aides in the realm of computing. Dartmouth students remained connected to their network and their Hanover peers via a terminal in their Jersey City residence, and they used a "roving" (portable) teletype to teach BASIC and "hands-on" computing at schools around the city.[60] The next issue of the newsletter again emphasized the ABC program with its report that a student had taken top honors in the prestigious John G. Kemeny Prize in Computing for Undergraduates "for his package of Computer-Assisted Instruction drill programs designed for use in the ABC Summer Project."[61] A year later, Kiewit boasted of its network of seventy-nine remote (off-campus) terminals, "including three in Jersey City operated by the Dartmouth College Urban Education Center."[62]

One might say that the college was rightfully proud of its support of the ABC program and how it nurtured connections between ABC students and computing; however, those reports nonetheless underscored the otherness and the difference of students who came from the ABC program. Moreover, because only some—but by no means all—black Dartmouth students arrived at the college via the ABC program, the emphasis on that program without reference to other African American students' paths to the college conflated the association of all black students with a program for disadvantaged students. Taking that one step further, one could have internalized the message that Dartmouth computing helped black students, all of whom were poor, and that all black students needed help with computing. In contrast, because there was never mention of special computing assistance or programs for white Dartmouth students, and because the Kiewit staff was predominantly white, one also could have internalized the message that whiteness equaled computing prowess.

Kiewit sent another mixed message on race and computing with its promotion of the homegrown programs RACECHECK and RACEMYTH. In the Computation Center's fall newsletter welcoming the class of 1974, the Kiewit staff touted the center's extensive library of 350 programs, with a special mention of GAMES including FOOTBALL, BLKJK, and POKER. They then mentioned the availability of RACECHECK, "which determines the probable racial group to which the user belongs," and RACEMYTH, "on the distribution of several physical characteristics which have often been used as indicators of racial differences."[63] The anthropology department, which had developed these programs, reported that RACECHECK was "designed to teach the difficulties and vagaries of racial classification. . . . It is fair to say that this program makes a persuasive attack on stereotypes of self (and other) racial identification."[64] I believe that the Kiewit staff had good intentions here. It seems they were encouraging the exploration of these programs by Dartmouth students to combat racial stereotypes.

The college had been experiencing a significant influx of black students and presumably saw RACECHECK and RACEMYTH as paths to decreasing racism; however, the Kiewit staff also seemed to miss the anthropology department's caution that "experience with [RACEMYTH] in an introductory course indicates it to belong more properly to advanced undergraduate work."[65] RACEMYTH required a deeper understanding of anthropological theories of race in order to be effectively utilized. In other words, encouraging a bunch of unprepared students to play with RACEMYTH, or even RACECHECK, might result in reinforcing racial stereotypes and bolstering racism. Moreover, when the Kiewit staff highlighted these programs, they called attention to the otherness of nonwhite students on campus.

The combination of computing technology with race was far from neutral. A program like RACEMYTH acquired seeming objectivity precisely because it was delivered in black and white on a teletype that was part of a substantial computing network. Yet the user's experience with such a program was highly dependent on what he brought to the teletype: his views, opinions, experiences, and biases.[66] The existence of programs like RACECHECK and RACEMYTH conveyed the message that race was something that could be studied through computing, and was therefore distinct from computing itself. That illusion of separation submerged the whiteness of Dartmouth computing.

Although Kiewit downplayed the white, heteronormative homogeneity of campus computing, it proudly located itself at the center of various local, regional, national, and international computing networks. Indeed, it called attention to its connections beyond the Hanover plain, perhaps purposefully to seem more diverse. Dartmouth emphasized its connections with other places through computing, locating the college in concentric communities that were both social and physical. In other words, these networks consisted of human resources such as people and the BASIC programs they wrote, as well as physical telephone lines, teletypes, and computers. From the time that Kemeny and Kurtz proposed their time-sharing network, they envisioned the college as a regional computing resource. Their 1963 application to the National Science Foundation suggested a "strong possibility that the Dartmouth Computation Center could be of significant service in bringing computing to some of the many small colleges" in New Hampshire and Vermont.[67] During the next decade, the college realized its vision, bringing computing to secondary schools and colleges throughout New Hampshire, Vermont, Maine, Massachusetts, Connecticut, and beyond. By 1971, Kiewit advertised itself as a "regional resource."[68] That regional network included thirty secondary schools and twenty colleges as regular users of Dartmouth time-sharing, with a total of seventy-nine remote terminals, including the three in Jersey City that were highlighted.[69] Over ten thousand non-Dartmouth students and educators regularly used the regional network. Chapter 3 analyzes the activities of those users.

Dartmouth produced and distributed multiple maps of its networks; these maps represented Hanover as a hub beyond its historical and geographical limitations sandwiched in the mountains along the Connecticut River in New Hampshire. Early in 1968, the *Kiewit Comments* circulated a map depicting the network's colleges and secondary schools.[70]

The map was striking because of the visual weight clustered around Dartmouth. The Kiewit logo, a stylized "K" made up of arrows, occupied the largest place, and it drew the eye of the viewer. The

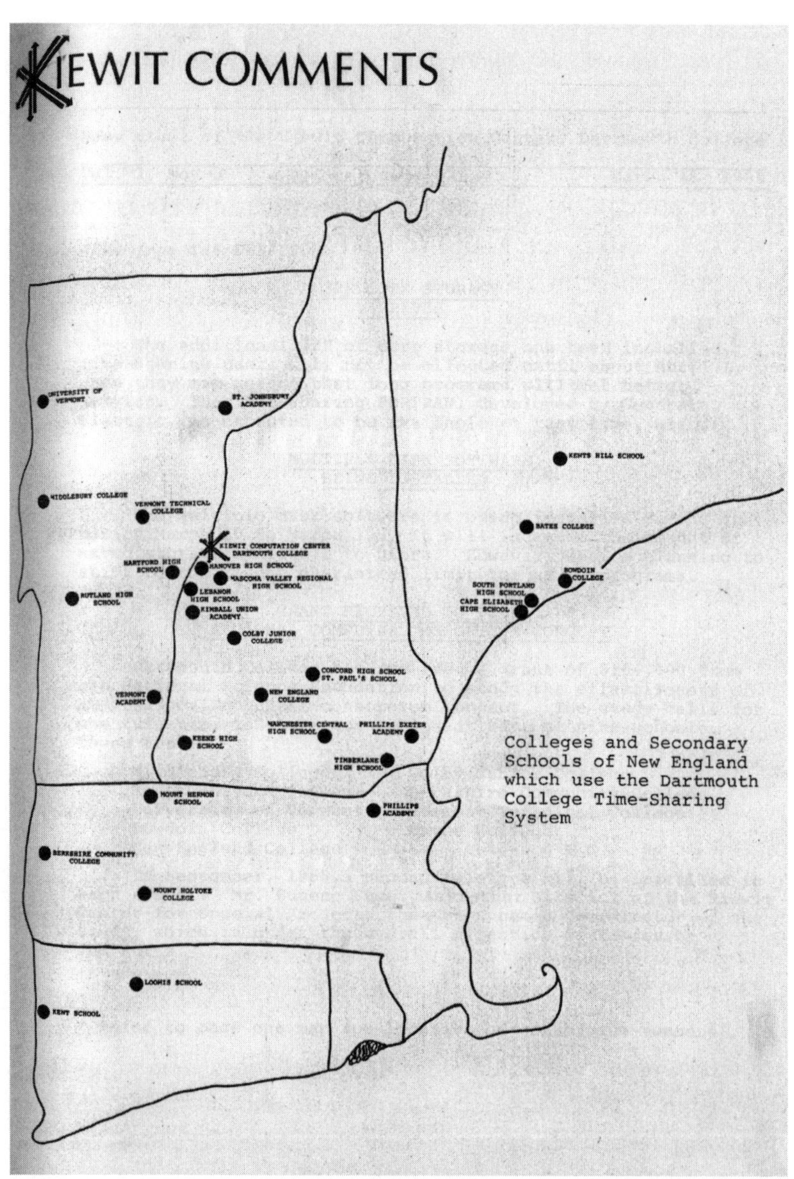

A 1968 map of the colleges and secondary schools using the Kiewit Network. Courtesy of Dartmouth College Library.

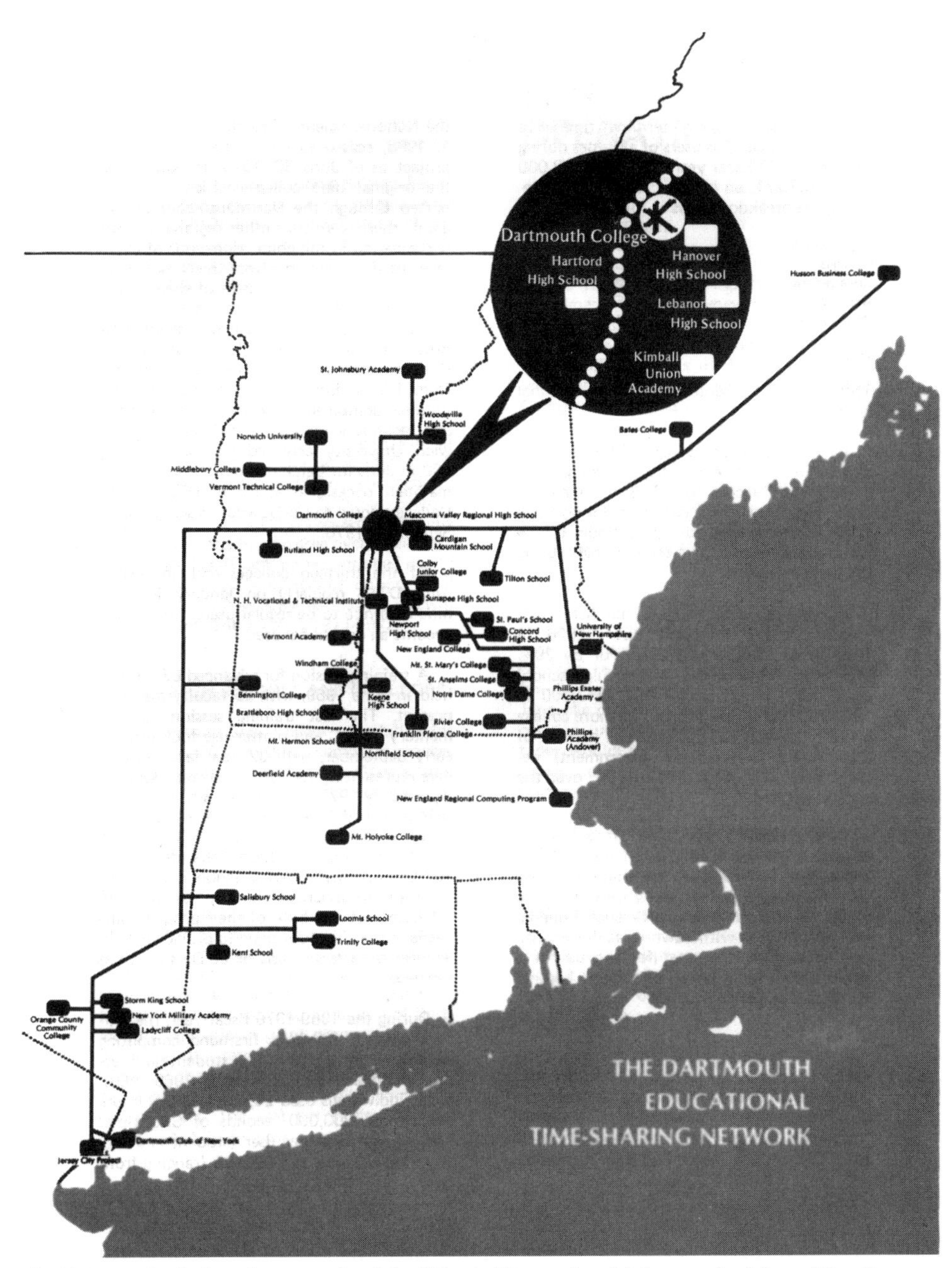

A 1971 map depicting the growth of the Kiewit Network, which stretched from New Jersey across New England and up into rural Maine. Courtesy of Dartmouth College Library.

logo, with its arrows pointing in opposite directions, immediately conveyed movement, expansion, and energy. Indeed, the logo embodied the potential and possibility of a network on which a student in Maine collaborated with a high school teacher in northwest Connecticut. Although the interstate highways I-89, I-91, I-93, and I-95 offered the fastest driving connections through New England during the 1960s, they crawled in comparison with a computing connection from one end of New England to another in minutes, or even seconds. The Kiewit Network offered a powerful imaginary of unity for people in remote areas, separated by mountains, valleys, and rivers that were often impassable in winter.

The appearance of this 1968 map both revealed and masked another feature of the Kiewit Network. There were more schools clustered around Hanover, giving Kiewit the added visual impression of centrality; however, that reflected the fact that long-distance telephone connections to Dartmouth were quite expensive, so schools that were farther away from Kiewit paid much more for their teletype connections. Finally, that 1968 map was also striking for what was not on it: the computing powerhouse of MIT. Although Boston, Cambridge, and the Route 128 corridor loomed large in the public awareness of computing during the 1960s and since, Dartmouth distinguished its computing—and its network—from its better-known neighbor to the southeast.

Dartmouth expanded and enhanced that hand-drawn 1968 map during the opening years of the 1970s. The map printed in the early 1970s Kiewit brochure and the 1969–1971 Kiewit biennial report retained the Kiewit arrow logo and the centrality of Dartmouth's campus.[71] This professionally produced map included lines to emphasize the connections between the various networked nodes and the Kiewit hub. The heavy, solid, tangible network lines, compared with the lightly dotted state demarcation lines, communicated the power of the network to overcome distance and traditional divisions. The black call-out circle focused viewer attention on a local network within the larger regional network. Indeed, Dartmouth's proximity to Hanover, Lebanon, and Hartford high schools, along with Kimball Union Academy (in Meriden, New Hampshire), meant that those students and teachers could easily visit Kiewit in person, which

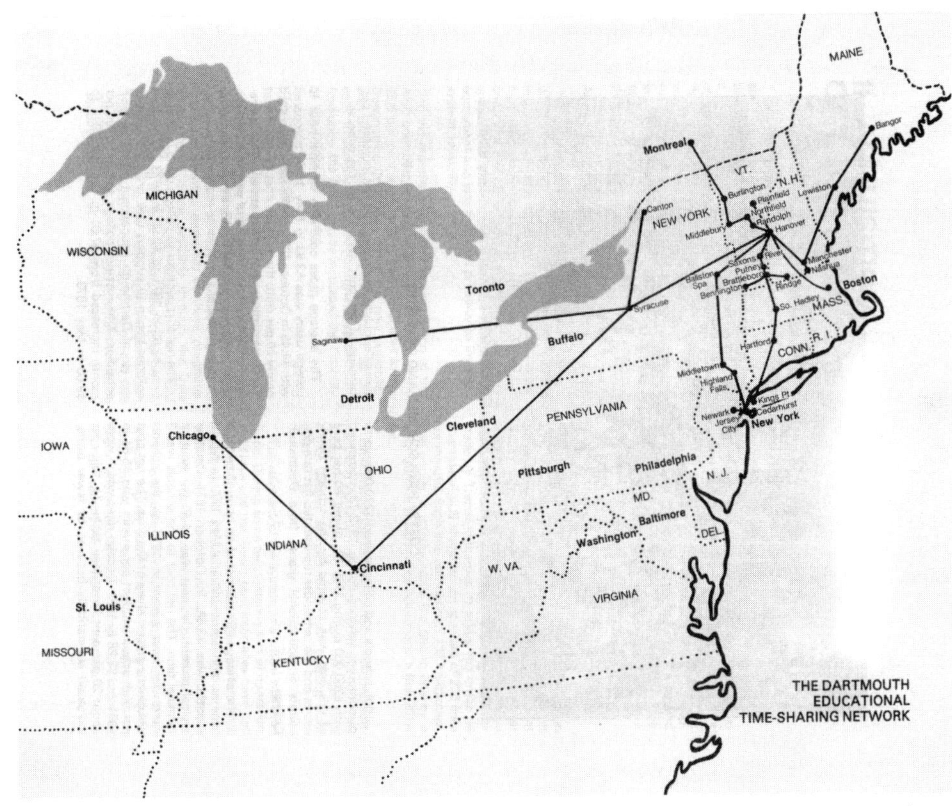

A 1973 map of the Kiewit Network, which had expanded beyond the northeastern United States to include Upstate New York, Ohio, Michigan, Chicago, and Canada. Courtesy of Dartmouth College Library.

afforded opportunities described in Chapter 3. The 1971 map pointed to the connection between Dartmouth and the Boston area via the New England Regional Computing Program; however, that was only one node among fifty. This map also showcased the expansion of the region to include New York City, New York State, and New Jersey, including the Jersey City project previously mentioned in conjunction with the ABC Program. The cluster of network locations around Dartmouth continued to be readily apparent; however, the 1971 map clearly proclaimed the network's reach with the line up to Husson Business College in Bangor, Maine, and down to the Dartmouth Club of New York and the Jersey City project.

The 1973 map attempted to convey the network's growing reach.[72] Kiewit traded a New England–focused map for one depicting the United States from Missouri to Maine, as far south as Kentucky and Virginia. Dartmouth also abandoned individual school names in favor of towns and cities. The resulting map delineated the growth of the network north to Montreal and west to Syracuse (New York) and Saginaw (Michigan), as well as Cincinnati (Ohio) and Chicago. What this map gained in geographic scope it lost in highlighting the network. Compared with the 1968 and 1971 maps, the 1973 map seemed to anticipate continued westward and southward expansion of the Kiewit Network; however, the small dots and little lines of the 1973 map made its purpose—and its network—unremarkable and undistinguishable. That the lines depicting the connections on the network (such as Cincinnati to Chicago) were not as heavy as the boundary line between the East Coast states and the Atlantic Ocean further diminished the visual impact of the network.

While the maps portrayed the Kiewit Network's growth, the Computation Center similarly publicized the national and global connections of its visitors and employees. Kiewit employee Samuel "Sammy" Karumba earned two mentions in the first five years of the *Comments*, first for his arrival, stating that he was originally from Nairobi, Kenya, and second for the birth of his daughter Kristina on April 6, 1970.[73] Other Kiewit newcomers hailed from Washington State, New York, Massachusetts, and California.[74] Four students from the Hatfield College of Technology in England spent six-month stints at Kiewit, and a faculty member from the University of Lyon in France moved to Hanover for a year with his wife and daughter to study computing at Dartmouth.[75] The newsletter mentioned visitors from the Pillsbury Company (headquartered in Minneapolis, Minnesota), which had bought its own GE time-sharing computer, and from the Tuskegee Institute in Alabama (which considered installing a version of Dartmouth time-sharing on its campus).[76] Finally, in 1969, Kiewit declared that "Xavier Aims to become 'Dartmouth of the Midwest,'" in an article describing the Ohio university's new time-sharing system, which was inspired by Dartmouth. The article emphasized Xavier's GE computer, how students enjoyed the freedom to use the teletypes as they wished, and how Kiewit director Kurtz had sent a congratulatory message to Xavier's president via—what else—teletype.[77] Kemeny, Kurtz, and their students were surely pleased that Xavier had emulated their beloved system.

Conclusion

When Greg Dobbs stepped out of his Butterfield dorm room in 1968 to stroll to Kiewit, he was in good company. He, like 80 percent of his student peers and 40 percent of Dartmouth faculty, embraced the college network's new personal and social computing.[78] Yet, during the 1960s, Dartmouth was almost exclusively white, male, and affluent. One might argue that because Dartmouth students were only men during the 1960s, the computing culture was not gendered masculine, macho or even heteronormative; rather, it was just a function of the fact that there were no students who were women. That misses the point of the processes by which gender and sexuality structure society. The centrality of football and other games to Dartmouth's computing culture, the subtle differentiation of the women who worked at Kiewit, the announcements of marriage and children in *Kiewit Comments*, and the courtship rituals demonstrating computing prowess all mutually constructed masculinity and computing at Dartmouth. During this formative first decade of Dartmouth computing, from 1964 through 1974, the Kiewit Center became increasingly central to social life on campus, and became the hub of a vibrant regional network. The users on that network were united by a common language, that of BASIC. We turn to the power and proliferation of BASIC from the Kiewit Network across the United States in Chapter 3.

3

Back to BASICs

BASIC (Beginners' All-purpose Symbolic Instruction Code) became the "lingua franca" of the people computing.[1] Students and educators alike praised it as easy to learn and conversational, with its syntax resembling simple words in the English language. During the 1960s, computer programming languages, including FORTRAN and COBOL, were still young. Prior to the development of languages like FORTRAN or COBOL, communicating with a computer entailed creating a step-by-step list of commands in machine language, which involved very specific tasks such as specifying a location in the computer's memory, storing a number in that location, or retrieving data from another location in memory. Those working with computers during the 1950s identified new programming languages, including FORTRAN and COBOL, as "higher level" because they bypassed the bit-by-bit, task-by-task nature of machine languages and instead permitted coders to call on library functions and preprogrammed routines. These included operations like computing sine or cosine, and statements including IF, READ, and PRINT. However, FORTRAN (which had been developed at IBM) and COBOL (supported by the federal government) were, from the start, intended for business, scientific, and engineering use. They were languages written by and for professionals. In fact, after Dartmouth rolled out BASIC (and in the decades since), many professionals and computing enthusiasts criticized BASIC for its simplicity, how it handled tasks, and the ways in which it did not maximize or fully utilize the power of the computer itself. However, those criticisms missed the point completely.[2]

Kemeny and Kurtz designed BASIC to make computing accessible to the widest possible range of users. BASIC maximized accessibility for most users, most of the time. BASIC was to be the language of the people, not the province of the professionals. Because Dartmouth College promoted BASIC and made it widely and freely available, BASIC users soon circumvented some of the problems of compatibility that generally plagued computing during the 1960s. During that decade, a program written for one computer model (such as an IBM) typically had to be rewritten to run on another computer (such as a GE). Moreover, programs were often created for business purposes, and were proprietary to those businesses, so they were rarely shared in the first place. By contrast, a program written in BASIC and published in a newsletter could, in most circumstances, be successfully run on a variety of computer models, with perhaps only minor changes.[3]

How did collaborating with a computer in BASIC compare with similar problem solving in FORTRAN? On February 5, 1967, under the headline "Computer Jumps to Ski Conclusions," the *New York Times* reported that a senior at Williams College had produced a program to score ski jump competitions. Normally, "it took more than three hours of concentrated figuring by a team of faculty and student statisticians" to calculate the results of any jumping meet, but the student promised the same results in thirty seconds with his FORTRAN program running on the college's IBM computer.[4] Dave Robinson, an instructor at Vermont Academy, just about fifty miles northeast of Williams, read the *New York Times* Sunday Sports Section with surprise and bemusement. Just the day before, Vermont Academy's Winter Carnival ski meet—ski jump, as well as cross country, downhill, and slalom—had been scored using a BASIC program that Robinson had written. Robinson used the Vermont Academy teletype connected to Dartmouth's time-sharing network. The most compelling contrast was that the Williams College student had spent fifty hours writing his ski jump scoring program in FORTRAN, whereas Robinson had spent only ten hours writing his comprehensive ski meet program in BASIC.[5]

This FORTRAN versus BASIC ski-scoring comparison illustrates two themes of this chapter. First, BASIC had been designed to be valuable to users: easy to learn, with a resemblance to English, amenable to operating on networks that used time-sharing, and both flexible and powerful enough to accommodate a wide range of creative computing impulses, not just those of business or academic professionals. Second, students, educators, and their networks popularized BASIC, making it the lingua franca of computing citizens. BASIC forms the heart of this chapter, just as it formed the through line of personal computing before personal computers. The first section presents the origins and syntax of BASIC, arguing that it was inseparable from networked computing. Time-sharing and BASIC together were intended to bring computing power to the people, and BASIC amplified that power by enabling users to easily share, swap, tweak, and build on one another's programs. Both time-sharing and BASIC created social communities around computing. The next section follows BASIC through its genesis at Dartmouth and in its spread around the Kiewit Network, the tens of secondary schools and colleges connected by Dartmouth time-sharing during the 1960s and 1970s. From 1964 to 1967, a handful of New England high schools tested the time-sharing waters, and their success with BASIC and computing informed the influential 1967 report of the President's Science Advisory Committee on computers in education; both Kemeny and Kurtz served on the panel on computers in higher education.[6] From 1967 through 1975, the users on the ever-expanding and changing Kiewit Network continued to embrace and popularize BASIC.

Meanwhile, an early BASIC adopter and aficionado, Bob Albrecht, shepherded the introduction of BASIC in Minnesota during the mid-1960s; he moved westward and became a BASIC proselytizer in the San Francisco Bay Area. From there, Albrecht championed other vectors through which BASIC spread: the popularity of Digital Equipment Corporation (DEC) and Hewlett-Packard (HP) minicomputers for time-sharing and BASIC, and the Huntington Project, an educational endeavor that created crowd-pleasing computer simulations in BASIC. In fact, DEC published Albrecht's *My Computer Likes Me When I Speak in BASIC*—a book that spread the gospel of BASIC via the more than 250,000 copies sold.[7] When Albrecht began publishing the *People's Computer Company*, a newsletter, in 1972, he continued to zealously endorse BASIC, and he promoted the Huntington Project educational modules in the first *People's Computer Company* newsletter.[8]

The Birth of BASIC

Kemeny and Kurtz's focus on users' needs prompted them to develop a new programming language: Beginners' All-purpose Symbolic Instruction Code, or BASIC. In fact, the development of BASIC and time-sharing at Dartmouth were intertwined from the beginning. The creation of BASIC stemmed from Kemeny and Kurtz's vision of computing for all their students. They envisioned that all students who completed a year of mathematics would be required to learn computing on the time-sharing system, a critical feature of their plans, and one that fostered a computing culture. To facilitate that goal, Kemeny and Kurtz wrote a language that could be learned within minutes and mastered within hours. BASIC enabled users to engage the computer for their own productivity and recreation. In short, Kemeny and Kurtz's focus on users inspired time-sharing and BASIC. In turn, BASIC and time-sharing launched personal computing at Dartmouth.

When he and Kurtz were planning for Dartmouth time-sharing in 1963, Kemeny contemplated the creation of a new programming language. He had been heartened by the ease with which his math major undergrads had learned the ALGOL programming language for the LGP-30, and their success with the Dartmouth Oversimplified Programming Experiment (DOPE), both described in Chapter 1.[9] Kemeny believed that truly widespread and quotidian computing at Dartmouth required a programming language that was "highly simplified" and "well-suited to the needs of the inexperienced programmer."[10] He benefited from the experience of Professor Richard Conway, who had recently implemented a streamlined programming language called CORC at Cornell University.[11] During the summer of 1963, Kemeny wrote a compiler for BASIC, a program that translated the user commands in BASIC to machine language for the mainframe computer to execute.[12] Although Kemeny and Kurtz were later celebrated as cocreators of BASIC, in 1964 Kemeny documented that he had designed the language and performed most of the programming and debugging himself ("with the aid of some of my colleagues").[13]

Although computing experts at the time and since criticized BASIC for its simplicity, Kemeny understood the trade-offs that a language like BASIC required, and he articulated a justification for each of those trade-offs. BASIC was to be a useful language for most people, in most of their programming endeavors, most of the time. I cannot emphasize enough that Kemeny wanted students and faculty to use the computer as much as possible—for their homework, for their research, and for their recreation—and he was convinced that computing would become a personal resource only if the programing language was approachable and memorable. He argued that it was worth shedding some of the "luxury items" of languages like FORTRAN and ALGOL to gain the incalculable benefits of a simple programming language.[14] He chose to simplify BASIC to the point that the Dartmouth Time-Sharing System (DTSS) compiled a BASIC program anew every single time the program was run. Why? Kemeny was adamant on this point: it gave the user the sense that the computer spoke BASIC and that the user was communicating directly with the computer.[15] Just as Kemeny and Kurtz had elevated user convenience in the design of Dartmouth time-sharing, so, too, did Kemeny uphold user convenience in the design of BASIC.

Many of the Dartmouth students who encountered BASIC for the first time during the fall of 1964 had probably never seen FORTRAN or ALGOL, but we can compare FORTRAN with BASIC to grasp just how refreshing BASIC was compared with the clunkiness of FORTRAN. A 1962 FORTRAN manual, *FORTRAN Autotester*, sold enough copies to merit a second printing in the summer of the following year, the season when Kemeny produced the BASIC compiler. Its authors wrote *FORTRAN Autotester* in a playful tone, "designed to emancipate the scientist and engineer from the need for the professional programmer."[16] Kemeny had a similar goal in mind. Yet *Autotester* ushered in the world of FORTRAN with a discussion of fixed point and floating point constants and variables, rules for vari-

Dartmouth students experimenting with teletypes at College Hall in 1965. Most students and educators on the DTSS (also known as the Kiewit Network) learned how to communicate, compute, and create on the network in the BASIC programming language. Photo by Adrian N. Bouchard, courtesy of Dartmouth College Library.

ables, and order of operations philosophy. Topics including "dimensions" and "arrays" were then introduced. Finally, after thirty-eight pages, the authors declared, "Up to this point, we hope you have learned the FORTRAN ingredients: symbols, card format, rules of constants, variables, etc. We now need to learn certain CONTROL statements which will give direction to the normal sequential order of events of a written program."[17] Only after thirty-eight pages did the *Autotester* authors address the creation of an actual program. This was only one approach to teaching and learning FORTRAN; however, the emphasis on the "ingredients" demonstrated that there was a lot about FORTRAN that an individual had to learn before he could even begin to think about composing a program.

When they publicized their fledgling system, Kemeny and Kurtz repeatedly highlighted how students learned to create meaningful and useful programs from a handful of elementary BASIC commands in a few hours.[18] One of the programs with which students could have experimented, and which illustrates seven of the nine elementary BASIC statements from 1964, was CONVRT.[19] CONVRT enabled the user to quickly convert measurements from the metric system (meters and centimeters) to the imperial system of feet and inches. The seven elementary BASIC commands were READ, LET, PRINT, IF-THEN (considered one command because the user had to have both in the program), GOTO, DATA, and END. (The remaining two elementary BASIC commands were FOR and NEXT.) The program would have been entered as follows (the numbers at the beginning of each line were part of the program and would have been typed onto the teletype):

```
100    READ M, C
110    LET M1=M+C/100
120    LET I=M1 * 39.37
130    LET F=INT(I/12)
140    LET I=I-12*F
150    PRINT M, "METERS,", C, "CENTIMETERS"
160    PRINT "CONVERTS TO"
170    IF F=0, THEN 190
180    PRINT F, "FEET,",
190    PRINT I, "INCHES"
200    PRINT
210    PRINT
220    PRINT
230    GOTO 100
240    DATA 1, 0, 0, 2.54, 3, 60, 2, 5
250    END
```

On line 100, the programmer used the READ command to instruct the computer to look for DATA (in this case, provided in line 240), and to label and use the first piece of data as M (for meters) and the second piece of data as C (for centimeters). Thus, when a user ran this program on the time-sharing system, the computer looked for the DATA line, and interpreted that M should be 1 and C should be 0.

After the computer completed the program for that first data set, it would move on to the next data set with M as 0 and C as 2.54, then M as 3 and C as 60, and finally M as 2 and C as 5. The user placed these four sets of data in one line, but he could have used four separate data statements (four separate lines) to provide this information to the computer. He could have provided forty values for the pair (M, C), or only one. Such was the flexibility of the program. The critical aspect of the DATA statement was simply the order in which the values appeared. With this program, the user instructed the computer to take the first value in the DATA statement as M(eters) and the next value in the DATA statement as C(entimeters). Thus, switching the 1 and the 0 in line 240 would have made a difference to the outcome of the program.

```
110        LET M1=M+C/100
```

With the 110 line, the programmer told the computer to recognize a new variable, called M1, and to make M1 equal to the value of M+C/100. This step converted the combination of meters and centimeters in the DATA statement into meters exclusively. (Dividing the centimeters number by 100 yields the equivalent number of meters.) For example, for the last DATA set of (M=2, C=5), the value for M1=2+5/100=2.05. Notably, the LET statement did not express algebraic equality. Rather, it commanded the computer to perform arithmetic operations and assign the results of those operations to a variable (in this case, M1).

```
120        LET I=M1 * 39.37
130        LET F= INT(I/12)
140        LET I=I-12*F
```

The user programmed the next three lines (120, 130, and 140) to perform the arithmetic of converting meters to inches and feet. The user assigned the variable F to represent feet, and the variable I to represent inches. There are 39.37 inches per one meter, so line 120 told the computer to multiply M1 by 39.37 to yield the total number of inches. In the case of the last DATA set (M=2, C=5), for which M1=2.05,

I inches = (2.05)*(39.37) = 80.7085. Line 130 told the computer to take the total number of inches and divide by 12, then take the whole number value of (I/12) for the number of feet. The INT function in BASIC commanded the computer to use only the integer (or whole number) part of a variable as specified. For example, if I = 80.7085, then INT(I/12) = INT(80.7085/12) = INT(6.7257083333) = 6 = F. Line 120 yielded the total number of inches (in this example, 80.7085). Line 130 computed the number of feet in that total number of inches (in this example, 6 feet). Line 140 told the computer how many inches remained after the number of feet had been calculated. That is, 80.7085 inches = 6 feet and 8.7085 inches.

```
150    PRINT M, "METERS,", C, "CENTIMETERS"
160    PRINT "CONVERTS TO"
```

The PRINT statements commanded the teletype to print both variables and specified text. In BASIC, anything appearing within quotation marks in a PRINT statement was then printed onto the teletype exactly as it appeared within the quotation marks. For this program, the teletype printed the value of M (meters) in the first column, followed by METERS, in the second column, then the value of C (centimeters) in the third column, followed by CENTIMETERS in the fourth column. The commas in the PRINT statement demarcated the columns. Line 160 commanded the teletype to print CONVERTS TO on the line below M METERS, C CENTIMETERS.

```
170    IF F = 0, THEN 190
180    PRINT F, "FEET,",
190    PRINT I, "INCHES"
```

With line 170, the user told the computer to check whether the value of F, feet, was equal to zero. If F = 0 for a particular DATA set, then the computer was to disregard line 180 and skip to line 190. That is, IF F = 0, THEN the computer should not print F FEET, and it should go to line 190 to print I INCHES. The IF-THEN combination in BASIC satisfied the critical requirement in computer programs to make decisions based on calculations already performed. In this example, the programmer employed the IF-THEN statement to save teletype time and printing by not printing 0 FEET if a particular conversion resulted in 0 feet and some inches.

```
200         PRINT
210         PRINT
220         PRINT
```

Lines 200–220 simply served to advance the teletype three lines (by printing blank lines) between one data set and the next data set. Here the user creatively employed the PRINT statement to separate out the results of the separate problems, making them easier to read.

```
230         GOTO 100
```

The programmer communicated to the computer that the program should be run for multiple data sets with line 230. Line 230 told the computer, "Now that you've run lines 100 through 220 for the first data set of M and C, GO back TO LINE 100 and run lines 100 through 220 for the next data set of M and C." The GOTO statement of line 230 enabled the user to input four data sets (four pairs of M and C) at once into the teletype, and to have the computer run the entire program and PRINT results for each data set. With CONVRT, the computer processed the commands from lines 100 through 220 for the value of M=1 and C=0, then again for M=0 and C=2.54, then again for M=3 and C=60, and finally for M=2 and C=5.

```
250         END
```

The END statement of line 250 indicated the end of the program, and with it the user communicated to the computer to stop computation.

After the user inputted the CONVRT program and typed RUN CONVRT, followed by RETURN on the teletype, the computer analyzed and ran the program. It then printed a line that contained the user number, the problem name, the date, and the time of day. Provided there were no errors in the program, the teletype then printed the answers or results of the program according to the PRINT statements within the program. Finally, the teletype printed a time statement that indicated the total computing time used by that programming run. If the time statement indicated 0 seconds, then the entire program required less than 0.5 seconds, a common occurrence.[20]

When a Dartmouth student ran CONVRT in the fall of 1964, the results printed on the teletype as follows:

USER NO. 999999	PROBLEM NAME: CONVRT	11 NOV. 1964	TIME: 15:05
1	METERS,	0	CENTIMETERS
CONVERTS TO			
3	FEET,	3.37	INCHES
0	METERS,	2.54	CENTIMETERS
CONVERTS TO			
.999998	INCHES		
3	METERS,	60	CENTIMETERS
CONVERTS TO			
11	FEET,	9.732	INCHES
2	METERS,	5	CENTIMETERS
CONVERTS TO			
6	FEET,	8.7085	INCHES
OUT OF DATA IN	100		
TIME:	0 SECS.		

Traveling BASICs

In 1964 Kemeny declared, "The heart of the time-sharing system is . . . BASIC," and he was right.[21] BASIC gave children, young adults, and their teachers the right tool for making the most of their time-sharing network. Indeed, Kemeny and Kurtz had anticipated a world in which their time-sharing system and BASIC would move beyond campus, starting with a local high school. Their 1963 proposal to the National Science Foundation (NSF) for a grant to incorporate computing into the college curriculum also pitched the idea of a teletypewriter at nearby Hanover High School.[22] Kemeny and Kurtz thought that some high school students might be interested enough in the new mode of computing to teach themselves BASIC and write some programs. The students in Hanover exceeded their expectations. They formed a computer club, which gained several hundred members over the course of the 1964–1965 school year.[23] The students impressed the college faculty with their creativity, and the Computation Center celebrated their accomplishments in its newsletter. Hanover students used the Dartmouth network to score a gymnastics tournament, ski events, and a debate tournament with 150 teams.[24] Students soon combined the novelty of computing with gaming. Twelve-year-old David Hornig programmed a version of solitaire called "Round the Clock" on the Kiewit Network as a summer project, and thirteen-year-old Julia Hawthorne developed a game of checkers "to apply the use of subscripts in BASIC."[25] Moreover, Hanover elementary students gained access to the time-sharing system, and a "bright fifth grader, working on his own," wrote a program for factoring integers.[26]

Kemeny fondly recalled providing a tour of Dartmouth time-sharing for the Hanover elementary students, including his son Robert, shortly after the system was set up.[27] He led them to the room full of teletypes upstairs in College Hall, and he demonstrated some programs. After that, "as a real treat," he showed them the GE computers in the basement. The following day, the teacher called Kemeny to report the students' enthusiasm for the tour, and to relay one student's memorable question. The girl had reported, "Well, I understood perfectly everything Mr. Kemeny did on the computer. . . . But then he took us downstairs into the basement and showed us a great big box that looked like a refrigerator. I never did find out what it was."[28]

Kemeny explained that initially he found the story funny, but it eventually signified an important insight for him: how users viewed and shaped the system. Kemeny concluded, "Ask yourself what you mean by a telephone," and that simple query belied the complexity of the concept he was expressing.[29] A college student calling home in 1964 experienced the entire infrastructure, technology, and human labor and maintenance of the vast telephone network simply by talking on a telephone. A telephone that was not connected to the network was effectively useless; however, a functional, connected telephone also very effectively masked the extensive behind-the-scenes wiring and work that enabled a successful call. Kemeny's implication was, "Ask yourself what you mean by a computer." He realized that for a Dartmouth time-sharing user, the interaction with the teletype often represented the entire network of telephone lines, computers, programs, and people that enabled his individualized computing experience. In other words, the teletype represented the entire Dartmouth network, with all of its technological (mainframe), communications (telephone), and social (people working behind the scenes, assisting and advising) resources. Kemeny keenly sensed that time-sharing fundamentally changed the nature of computing.

Kemeny and Kurtz may have thought that programming on a time-sharing system was by itself compelling enough to generate such interest; however, the students of Hanover received encouragement and stimulation from their geographic and social circumstances. Hanover High was located at the southeastern edge of Dartmouth's campus, a few hundred yards from the college football stadium. Hanover was (and is) a college town: the high school students saw Dartmouth students all around them, and the college employed many of the Hanover High parents. The teletypewriter at their high school was just the beginning of access for interested Hanover High students. They could walk a half mile up Lebanon Street and then South Main Street to use the Dartmouth teletypes in College Hall; indeed, one member of the Dartmouth class of 1968 received assistance from an eager secondary school student. Francis Marzoni recalled asking for programming help from "a 12 year old sitting next to me who was typing up a storm . . . trying to figure out how far Earth [would] be from Mars at the Vernal Equinox."[30] When a Hanover High student needed help working through some aspect of a program, the student could seek out a Dartmouth student or faculty member, possibly a parent. Indeed, during 1965 and 1966, time-sharing teletypes appeared in homes around Hanover. A special correspondent to the prestigious British science journal *Nature* reported in 1966 on these computers in the home, and the striking informality of the Dartmouth system.[31] Thus, students at Hanover High who were interested in computing did not have to rely on a teacher as their only computing resource. Rather, they learned to rely on each other, and on the computing community of Dartmouth students and faculty that enveloped them. When they found their high school teletype occupied, they could seek out the sociability of the college teletype room or perhaps a residential teletype. Hanover's access to computing resources (connected via relatively inexpensive local telephone lines) and computing expertise produced a strong experience of social computing for the high school students.

Enchanted with the success of the Hanover experiment, Kemeny and Kurtz readily acquiesced when other high schools requested time-sharing access; the college even provided educational grants to cover the costs of teletype rentals and long-distance telephone charges for some of the schools.[32] During 1965 through 1967, eight more high schools connected to Dartmouth (see Table 3.1).[33] That these schools asked Dartmouth for time-sharing reflected the interest and motivation of some individual or group at each school, thereby ensuring that interested students had a source of support and engagement for their computing endeavors. For example, the administration at Mount Hermon School arranged for four hundred students (half of the student population) to receive training in BASIC on the teletype during a single fall semester.[34] At Mascoma Valley, a small rural school, the enthusiastic student response convinced the school district to arrange for its own teletype rental and long-distance connection to Dartmouth.[35] As Dartmouth's time-sharing network grew, so did the number of individuals exposed to BASIC. Kemeny and Kurtz happily received these reports. They were fostering the widespread use and sharing of the BASIC language and BASIC programs.

Dartmouth enjoyed a particularly strong relationship with Phillips Exeter Academy, brokered by the strong support of its mathematics instructor John Warren. This prestigious high school in southeastern New Hampshire (about one hundred miles from Dartmouth), often known simply as Exeter, added a teletype to the

Table 3.1 Schools Connected to the Dartmouth Time-Sharing System by 1967

School	State	Public or private
Hanover High School	New Hampshire	Public
Phillips Exeter Academy	New Hampshire	Private
Phillips Andover Academy	Massachusetts	Private
Mascoma Valley Regional High School	New Hampshire	Public
Mount Hermon School	Massachusetts	Private
Vermont Academy	Vermont	Private
Kimball Union Academy	New Hampshire	Private
The Holderness School	New Hampshire	Private
St. Paul's School	New Hampshire	Private

Source: John G. Kemeny and Thomas E. Kurtz, *The Dartmouth Time-Sharing Computing System: Final Report*, 24-26, Box 8, Stephen Garland Papers, ML-101, Dartmouth College Rauner Special Collections Library; also available at http://eric.ed.gov/?id=ED024602.

Dartmouth network in January 1965, just a few months after Hanover High School had connected. Exeter enrolled just under eight hundred students for its four-year program, and in 1968, 98 percent of those students continued to a four-year college or university. That year, only 62 percent of Hanover High's nearly seven hundred high school students continued to a four-year college, while only 35 percent of Lebanon High's students advanced.[36] Understanding those percentages means keeping a few things in mind: Hanover and Lebanon were both public high schools, attended by both girls and boys. However, during the 1960s, it was still far less common for girls to continue to college than boys.[37] Exeter was both private and all-male, but its 98 percent college-bound rate also reflected the relative wealth and status of most Exeter students. Dartmouth and Exeter nurtured their computing ties even before time-sharing had been implemented in Hanover. In December 1963, Exeter instructor Warren helped articulate how computing should be integrated in the Dartmouth math curriculum, and he then spent the summer of 1964 in Hanover helping with the launch of Dartmouth time-sharing.[38] A little over a year after Exeter joined the Kiewit Network, a student reported that computing had "become a focal point for a large number of stu-

Students working on a time-sharing teletype terminal. Students in public and private schools across New England accessed computing through the Kiewit Network. Courtesy of the Computer History Museum.

dents . . . writing their own programs."[39] He also extolled the "virtue of BASIC . . . that just about anyone can learn to use it efficiently and effectively after a few hours of instruction," and he estimated that "students and faculty have written several thousand programs."[40] At Exeter, games were especially popular, with programs for bingo and bridge, poker and roulette, and baseball, football, and golf. Such games were not mere "idle pastimes" declared the author, since Exeter students demonstrated "considerable skills and talents" in creating them.[41] When the Kiewit Network welcomed nearly 20 more secondary schools in 1967, the Exeter instructor Warren trained them.

Encouraged by these high school successes, Kemeny and Kurtz formalized their support of high school computing by securing an NSF grant for a secondary school time-sharing network. They initially pitched the project for two years; however, some budgetary miscalculations enabled the project to continue for a third year.[42] Dartmouth applied the grant to teletype rentals and long-distance telephone lines for the schools in the network, and to hire a coordinator who organized training, newsletters, and teacher and student support for the network schools. The project's promised deliverables were some "topic outlines" for high school teachers on how to use computing in various high school courses.[43] Thus the secondary school students and teachers enjoyed freedom in exploring computing for their own purposes. Although Kurtz highlighted these curriculum units as a "primary goal," subsequent reporting by Kurtz and the project coordinators conveyed the sense that they simply wanted to put time-sharing and BASIC into as many hands as possible and observe the results.

The Dartmouth Secondary School Project connected the students and educators in eighteen high schools from Connecticut to Maine, ranging from those in rural farming communities to those at elite private schools (see Table 3.2). The project organizers emphasized the diversity that they noticed: less than half of the public school students attended a four-year college after graduation, while nearly all of the private school students continued to college. The project reporters, including Kurtz, John Nevison, and Jean Danver, did not comment on the racial or socioeconomic composition of the participating schools; however, other contemporaneous accounts offered a window on those characteristics. In 1969 Dartmouth president John Sloan Dickey described both the "poverty problems" and the "issue of race relations" in rural New Hampshire and Vermont, where schools such as Hartford, Keene, Lebanon, and Rutland were located.[44] During the previous handful of years, Dickey had collaborated with the leaders of the elite private academies to diversify their student populations through the ABC (A Better Chance) Project, with the goal of diversifying college populations. To some extent, the dramatically varying percentages of Secondary School Project graduates who continued to a four-year college reflected the dramatic

Table 3.2 Schools Participating in the Dartmouth Secondary School Project during 1967–1968

School	State	Public or private	Number of students	Students to four-year college (%)
Cape Elizabeth High School	Maine	Public	573	48
Concord High School	New Hampshire	Public	1,478	38
Hartford High School	Vermont	Public	516	24
Hanover High School	New Hampshire	Public	679	62
Keene High School	New Hampshire	Public	1,415	33
Lebanon High School	New Hampshire	Public	536	35
Loomis School	Connecticut	Private	444	100
Manchester Central High School	New Hampshire	Public	1,616	42
Mascoma Valley Regional High School	New Hampshire	Public	456	30
Mount Hermon High School	Massachusetts	Private	620	99
Phillips Andover Academy	Massachusetts	Private	860	95
Phillips Exeter Academy	New Hampshire	Private	789	98
Rutland High School	Vermont	Public	1,063	33
St. Johnsbury Academy	Vermont	Private/Public*	701	32
St. Paul's School	New Hampshire	Private	458	100
South Portland High School	Maine	Public	1,700	50
Timberlane High School	New Hampshire	Public	830	34
Vermont Academy	Vermont	Private	217	95

* The interim report explains (on pages 2–3): "It should be noted that St. Johnsbury Academy will be considered a public school in this report because it serves as the sole secondary school for the town of St. Johnsbury and the composition of the student body resemebles that of other public schools. Also, it used the teletype from 8 a.m. to 4 p.m., Monday through Friday. (Private schools used the teletype from 8 a.m. to 8p.m., Monday through Saturday.)"

Source: Thomas E. Kurtz, *Demonstration and Experimentation in Computer Training and Use in Secondary Schools: Interim Report, Activities and Accomplishments of the First Year*, October 1968, Dartmouth College History Collection, Dartmouth College Rauner Special Collections Library.

socioeconomic differences among the rural high schools and the prestigious private academies.

By all accounts, the students loved BASIC computing on the Kiewit Network. Nevison's report to the NSF on the first year of the project centered on the "lone user at the teletype," whom he regarded as "our most important teacher this year."[45] Nevison celebrated this "lone figure of a high-school student seated in front of a teletype terminal" as "the heart and core" of the project, while invoking Dartmouth's motto of a "lone voice crying in the wilderness."[46] The secondary school project really was in the hands of the students that first year; only one teacher at each school had received some BASIC training, and the students were encouraged to explore time-sharing, BASIC, and programming for their own purposes.[47]

The high school students "saturated" the available teletypes for up to twelve hours a day, six days a week.[48] The two South Portland High School teachers using the teletypes reported that they began going to the school on Saturdays to use them because students monopolized the teletypes during the week; however, after a student spotted the teachers at school one Saturday, the students clamored for—and received—Saturday access too.[49] Some young men at Phillips Exeter Academy woke at 4:00 in the morning to use the teletype.[50] The project records manifested that the teletypes were so popular that schools had to regulate access. Nevison proffered the Mount Hermon School as a model: students signed up for fifteen-minute slots, and they were allowed a maximum of two (nonconsecutive) slots per day. Such policies prevented the small group of extremely zealous computing enthusiasts at each school—the "hard core users"—from dominating the teletype.[51]

How did the students experience personal and social computing? They enthusiastically contributed to and read their student gossip file, "a commonly available file in the computer to ask and answer questions of students at other schools." This file could be accessed, modified, and read by any student on a teletype in the secondary school network. The students shared and consumed news—"gossip"—about what was happening at their schools, creating social connections from Connecticut to Maine. Indeed, the students protected the integrity of the gossip file as a point of pride. As Nevison relayed, "Any prankster at the school could have destroyed it. . . . Yet, because the students knew it was theirs, it was successfully used all year with only a few minor mishaps."[52]

These students also created an imaginative range of programs, including many games. Some of the programs performed mathematic tasks such as solving systems of algebraic equations, factoring polynomials with rational roots, and calculating the area under a curve. Other programs reflected curiosity and creativity: a program producing haiku poems, a program for preparing one's federal and state income taxes, and programs to score sailboat races, organize basketball practices, and lay out the school newspaper. The games and simulations included horse racing, roulette, battleship, poker, basketball, bowling, hockey, and soccer.[53]

Indeed, the Dartmouth project's encouragement of game creation and game playing set it apart from many other educational computing endeavors of the decade. Kemeny and Kurtz praised gaming for drawing users into computing and for fostering both comfort and curiosity.[54] That attitude also pervaded the Secondary School Project. Nevison noted, "Among the most interesting and complex programs written by students in the project were many 'games.' Most were written for fun; some were written to be used in science classes."[55] One student wrote a BASIC version of the now-iconic game *Spacewar!*, first developed only a few years earlier on a DEC PDP-1 at MIT.[56] Another student combined computing and art with a digital re-creation of Robert Indiana's iconic LOVE sculpture.[57]

The students expressed their fondness for computing when evaluating the Secondary School Project. One student reported, "I have used the computer for entertainment. I find it very intriguing but hard to write a useful program. I have also played most of the games stored in the Dartmouth Center."[58] Another reflected, "If the terminal were removed, I as well as many others would try to have it brought back, because there are a great many who almost depend on this machine to do a lot of their assignments."[59] Of course, not all of the students were enamored of the teletype. One asserted, "I find the computer a waste of time. Since I know I will never use it after I get out of school, I don't see the point of using it."[60] This student likely could avoid further computing, however, since most of the schools did not have any kind of computing requirement.

The Dartmouth network of computing high schools—and the importance of BASIC—garnered substantial local and regional attention. The *Dartmouth Alumni Magazine* reported on the project, and the Kiewit staff provided regular newsletter updates on the activities of these students and teachers.[61] The Loomis School (now Loomis Chaffee) and Mount Hermon proudly shared the news of their computing activities in their school bulletins; their students had been "introduced . . . to the marvels of the computer, not only for work, but for fun and games as well."[62] In July 1967, a dozen New England newspapers—including the widely circulated *Boston Globe*—announced the commencement of the project at the end of the summer, and cited what had already been accomplished at schools such as Hanover High and Phillips Exeter. "One group prepared a program in BASIC language which plays chess. . . . Another student developed a program that composes Japanese haiku poetry."[63]

These articles invoked a strong sense of place, power, and connection. The New England press emphasized the locations of all the networked schools, as well as the fact that they would all be "linked."[64] Students gained "quick and easy access to a multimillion dollar computer by simply completing a telephone call," and up to two hundred people could use that multimillion-dollar computer at the same time—including Dartmouth students and faculty.[65] Such reporting conveyed the ways in which the Dartmouth network brought individuals, schools, and the region together around computing while encouraging personal computing use: "We start from the premise that in the lifetime of today's students the use of computers will become as much a part of everyday life as the telephone or automobile," explained Kurtz.[66] Through the Dartmouth network, thousands of students personalized their computing, and their communities previewed the possibilities of computing connections.

Computing Made Masculine on the Kiewit Network

Who were these students? Mostly boys. Among the secondary schools that composed the Dartmouth network prior to 1967, seven (of the nine) were private and all-male. During the 1967–1970 project, all of the private schools in the network were boys-only.[67] Furthermore, the boys at the private schools received teletype access for seventy-two hours per week, compared with only forty hours per week for public school students.[68] At the public schools, the computing enthusiasts also tended to be boys. Of thirteen student program authors acknowledged in two bulletins, twelve were male.[69] Of the eight Hanover High students who completed summer computing projects in 1967, seven had masculine names, and when the school had a computing contest in 1971, all seven of the entrants again had masculine names.[70] The formal reports on the project devoted significant attention to "the students," but there was no mention of their gender. Entries for the Kiewit Cup programming contest, in which students could enter their BASIC programs to receive accolades and a small prize, were listed by school or by grade. Perhaps this omission reflected the assumption that these students were mostly young men. And here we come to a crucial point.

Although the educational and career options and expectations for young women would change dramatically during the decade after the Secondary School Project, most of the girls attending these public schools in New England in the late 1960s expected to marry young and live as homemakers. Indeed, circa 1970, high school boys were still enrolling in many more math and science classes than high school girls.[71] In the 1960s, girls typically were not encouraged in high school mathematics, and because computing was closely associated with mathematics, girls were not encouraged to compute. In fact, although BASIC was celebrated as easy to learn and similar to English, it was also touted for being very algorithmic—that is, very mathematical. At Dartmouth, BASIC had been the product of math professors and was taught in math courses, and when time-sharing and BASIC moved out into the world, they carried that association.

The project teachers observed that a small group of computing zealots emerged at each school, and a now-familiar masculine computing pattern emerged.[72] The young men who displayed great interest in computing were encouraged, and their disruptive behavior was only mildly chastised. In short, the Dartmouth network encouraged the peculiarly masculine form of computing in which

A teacher explaining BASIC at a teletype terminal. Although girls accessed the Dartmouth network and eagerly computed, many more boys gained access through their enrollment in single-sex private schools. Courtesy of the Computer History Museum.

computing prowess outweighed behavior that was problematic to the community, behavior dismissed as youthful hijinks. In some ways, this was a reciprocal and symbiotic relationship. Kemeny and Kurtz repeatedly and vocally praised Dartmouth undergraduates and secondary school students for their computing efforts. What the students accomplished made Kemeny and Kurtz look good, but the students undoubtedly enjoyed the attention. For example, *Kiewit Comments* highlighted the secondary school students as "adventuresome" in their use of new programming languages, in the same issue that a Hanover High student bragged about the diversity of computing efforts at his school.[73] Kiewit staff acknowledged another Hanover student for his LISP program and as a "young assistant" for his work on a Dartmouth computing grant funded by the Department of Defense.[74] Yet, Dartmouth computing citizens also complained about the "breach of user etiquette among our younger users," who caused interruption and annoyance when they prank-called teletypes.[75] Dartmouth computing celebrated and castigated these young users, but the emphasis on the celebration engendered an association of computing prowess with adolescent masculinity.

Nonetheless, the teachers also noticed that a steady group of students used the teletypes occasionally, but with interest. In his final summary on the project, Nevison reported there were an average of eighty-three users per month at the public school teletypes, and about 124 users per month at the private school teletypes.[76] Nevison and Jean Danver, who coordinated the project during the second year, reported at the Spring Joint Computer Conference in 1969 that about 25 percent of the high school students used the teletypes at least occasionally, emphasizing that "this figure is extraordinary when you consider that, for most of the schools, there was only one person to teach them the BASIC language."[77] Thus, thousands of students (and teachers) on the Dartmouth network were computing.

The Secondary School Project included one other—especially overlooked—group of users: the teachers. Here, the intersection of gender and role is more complex. Although we think of teaching as a historically feminized profession, almost all the teachers advising the secondary school project were, in fact, men.[78] But two of the teachers routinely highlighted for their excellence as computing advisers were women: Mary Hutchins at Hanover High School and Ann Waterhouse at South Portland. Some students bragged about how much faster they mastered computing than their teachers, but other students viewed their computing advisers as wise mentors.[79] The teacher was, of course, a computing citizen herself—she had to learn BASIC and then teach it to her students, and she provided a sounding board for questions and problems.

If the students and their teachers embodied the nodes of the secondary school network, then Dartmouth provided the hub. The project coordinator resided at Dartmouth, and the coordinator attempted to support each school's computing endeavors and to facilitate communication between the schools. Although the teacher gossip file failed, the student gossip file connected high schoolers who were hundreds of miles apart. Dartmouth published a biweekly project newsletter that featured relevant system updates, as well as student and teacher programs.[80] The college coordinated the Kiewit Cup programming competition to encourage creative programming by individuals and schools.[81] Dartmouth organized some in-person knowledge sharing via a few weekend conferences and school visits by the coordinator, but the college played an arguably more important role as a help desk for the project students and educators.

In the early months of the project, Nevison and the Dartmouth students maintaining the DTSS at Kiewit Computation Center realized that many problems with network school service originated with telephone connections and telephone service. Over time, rather than having each project school contact the telephone company, Dartmouth advised all the project schools to contact someone at Kiewit directly. Nevison recognized that the telephone company responded much faster to Dartmouth (as a large client) than to any of the project schools. Shortly after that, an individual at Kiewit was designated the support person for the Secondary School Project.[82]

Dartmouth also provided the impetus for the creation and dispersal of numerous innovative computing approaches. Kemeny and Kurtz's encouragement of the writing and playing of games inspired many students and their Kiewit Cup entries. The project's requirement to produce curriculum materials compelled teachers. Dartmouth publicized these endeavors through the project newsletter, which circulated to over one hundred schools within a year, and by formally publishing one textbook, thirty-eight topic outlines, three teacher's guides, and five booklets for students, including *BASIC in Ten Minutes a Day*.[83] Dartmouth created and circulated a much-needed educational resource: concrete instructions for teachers and students on how to use the time-sharing systems that were proliferating in their schools. The outlines included titles such as *Four Classes with Fourth Graders*, *Solutions of Simultaneous Linear Equations*, *Ninth Grade Word Problems*, *Genetics of the Fruit Fly*, *The Use of the Computer in Air Pollution Study*, *Numerical Integration*, and *Free Falling Bodies and Projectile Motion*.[84] One particularly devoted teacher, G. Albert Higgins Jr., prepared an entire textbook, titled *The Elementary Functions: An Algorithmic Approach*, in which students spent at least an hour per week using the teletype.[85] Dartmouth's development of these resources coincided with the emergence of more affordable time-sharing minicomputers from companies including DEC and HP.

The Kiewit Network Grows

Dartmouth mirrored its Secondary School Project with a College Consortium, funded by an NSF grant during 1968–1969 and 1969–1970. The colleges, listed in Table 3.3, included Bates, Bowdoin, Middlebury, and Mount Holyoke. After Dartmouth had implemented time-sharing, representatives from numerous other colleges and universities visited Hanover to observe the system, and many requested access.[86] The university merely needed a teletype (which it probably already had somewhere on campus or could easily rent) and a (long-distance) telephone line between the requesting university and Dartmouth. Kemeny and Kurtz happily agreed to these requests for access, viewing it as missionary work to spread the gospel of BASIC and time-sharing.[87] Moreover, during the 1950s, Dartmouth had benefited from its participation in the cooperative New England Computation Center at MIT (the reason for Kurtz's long computing-commuting treks). Thus, Kemeny and Kurtz envisioned Dartmouth as a time-sharing resource for other New England colleges. In their 1963 proposal to the NSF, they projected "bringing computing to some of the many small colleges" in New Hampshire and Vermont.[88]

Table 3.3 Colleges Participating in the Dartmouth College Regional College Consortium

College	State	Approximate enrollment	Men (%)	Women (%)
Bates College	Maine	1,050	55	45
Berkshire Community College	Massachusetts	1,000	Not reported	Not reported
Bowdoin College	Maine	1,000	100	0
Colby Junior College	New Hampshire	600	0	100
Middlebury College	Vermont	1,600	60	40
Mount Holyoke College	Massachusetts	1,800	0	100
New England College	New Hampshire	1,000	75	25
Norwich University	Vermont	1,200	100	0
Vermont Technical College	Vermont	450	Not reported	Not reported

Sources: Thomas E. Kurtz, *Interim Report of July 1969 on the Dartmouth College Regional College Consortium*, Box 7977, Records of Dartmouth College Computing Services, Dartmouth College Rauner Special Collections Library; Thomas E. Kurtz, *Dartmouth College Regional College Consortium Final Report to the National Science Foundation 1970*, Box 7977, Records of Dartmouth College Computing Services, Dartmouth College Rauner Special Collections Library. During 1968–1969, the consortium also included the University of Vermont. The *Interim Report* tallied 6,483 users as a preliminary total for the consortium for 1968–1969.

In their 1967 final report on their NSF grant to integrate computing in Dartmouth's curriculum, they suggested a Regional Computation Center. They secured a two-year NSF grant to support the College Consortium, with the nominal goal of determining whether a time-sharing consortium met the computing needs of the involved institutions.[89]

The grant basically paid for teletype rentals and telephone costs, but the consortium was very loosely organized, especially in comparison with the Secondary School Project. The consortium grant required no formal deliverables from the member colleges; there was no Kiewit Cup, no requirement for faculty or student creation of BASIC programs. An attempted faculty newsletter failed. Many of the faculty members who received training over the summer had lost interest by the time they received teletype access.[90]

Despite its loose organization, the consortium illuminated the value of the Dartmouth network as a commons, a communal space whose sum equaled more than its teletype, telephone, and computer parts. Kurtz declared, "The greatest achievement of the entire experiment is the creation of a common system for inter-disciplinary communications by professors in the northern New England area."[91] Kurtz explained that the common resource of the time-sharing system, together with its associated trainings and seminars, shared program library, and personal visits among campuses, contributed to communication, and to the commons. BASIC provided the language for this computing commons. Kurtz emphasized, "This communication is not limited to terminals,"[92] and he had identified a nascent feature of the Dartmouth network: it encouraged personal connections and information sharing both on and beyond the computer. Users on the Dartmouth network created computing that was not just for problem solving or information processing; it was also for entertainment, recreation, personal satisfaction, gossip, and information sharing.

Thousands of students at these nine colleges and universities used DTSS and BASIC personally, socially, productively, and creatively. Notably, the women at Mount Holyoke corresponded with Dartmouth men (perhaps to arrange Winter Carnival dates, or keep in touch with boyfriends) over the network.[93] David Ziegler, a member of Dartmouth's class of 1968, relayed the following story that illuminated another use of the network:

> In the fall of 1967 I was dating Myra, the person who has now been my wife for the last forty five years. Myra was a senior at Mt. Holyoke. She was taking a course in BASIC from South Hadley but using the Dartmouth computer. At the time long distance telephone calls were expensive. In 1967 dollars, it would cost me about $2.50 to have a three minute conversation with her on the dorm pay phone. That would translate to something like $15.00 today. Myra told me that she had a half hour time slot and there were a number of other women taking the same course. I became curious as to who was paying all of these long distance phone bills, so I asked her. She told me that somehow Mt. Holyoke had a local Hanover phone line in South Hadley. We then wondered if Myra could use that line to call my dorm. It turns out that she could. So, for the rest of her semester, she would get her work done on the computer in fifteen or twenty minutes and then call me, and we courted that way. We got married by a Notary Public upstairs in the firehouse in Hanover after English class on May 27, 1968.[94]

Although the consortium formally disbanded when the grant ended, computing persisted on many of the campuses. Bowdoin and Norwich purchased their own time-sharing systems, and Colby then accessed time-sharing via Bowdoin's DEC PDP-10 (instead of Dartmouth's GE computers) to reduce telephone costs.[95] The other six consortium members remained connected to the Kiewit Network; indeed, by April 1971, Dartmouth's Kiewit Network had expanded to include thirty high schools and twenty colleges in New England, New York, and New Jersey.[96] The college reported over thirteen thousand users, with only three thousand at Dartmouth.[97]

The BASIC Missionary: Bob Albrecht

Kemeny and Kurtz spread BASIC primarily with well-ordered projects supported by the NSF, but Bob Albrecht employed a grassroots approach. An early advocate of computing for kids, Albrecht learned about BASIC and time-sharing while teaching at the University of Minnesota Laboratory High School. In the early 1960s, Albrecht toiled as an analyst for the Control Data Corporation in Colorado when a serendipitous invitation transformed his corporate path. Albrecht agreed to speak with a group of high school students about his work; when he casually asked them if any were interested in learning how to program, all thirty-plus students enthusiastically responded yes. Albrecht taught them, and he persuaded them to teach their peers. He was impressed and inspired by their zeal for—and ability to learn—computing, and so was Control Data. The corporation supported him as he reoriented his job to travel around the country demonstrating computing and teaching programming to high school students, as well as showing off the kids' programming prowess at national computing and education conferences.[98]

Albrecht moved his home base to Minnesota, where Control Data had its headquarters, and there he encountered BASIC. He taught programming at the University of Minnesota Laboratory High School, which by 1965 featured a computing connection to the DTSS, and where the young mathematics teachers praised BASIC. Albrecht converted from FORTRAN to BASIC and never looked back. He felt that FORTRAN was a terrible language in general, never mind for educational purposes.[99] He established the Society to Help Abolish FORTRAN Teaching, known as SHAFT, and he distributed SHAFT pins and pamphlets at schools and conferences over the course of his travels.

Albrecht attended regional and national meetings of the National Council of Teachers of Mathematics (NCTM), and he extolled the virtues of BASIC and time-sharing to everyone. When the NCTM appointed a Computer-Oriented Mathematics Committee, Albrecht was on it. The committee published the book *Computer Facilities for Mathematics Instruction*, which was an introduction to computing, and a book on BASIC.[100] In fact, the NCTM report on *Computer Facilities* praised time-sharing as the "least expensive and most efficient way to use computers as instructional tools in teaching mathematics."[101] Moreover, the NCTM's publication of a book on BASIC de facto endorsed BASIC as the computing language of choice for its students and educators.

In the meantime, Albrecht fled the icy and buttoned-up environment of Minnesota and Control Data for the freer and warmer climate of San Francisco, where he evangelized BASIC and kids computing with his roadshow and his publications. In California, Albrecht intertwined his success with that of DEC. Throughout the 1960s, Albrecht traveled around the United States, spreading the gospel of BASIC.[102] He continued this in California during the early 1970s, now with a computer in the back of his van. He had received a DEC PDP-8 minicomputer with a teletype in a mutually beneficial exchange. This DEC minicomputer occupied the space of a bookshelf, which was positively portable compared with room-dominating mainframes. DEC salespeople traveled with the PDPs in their trunks for demonstrations, and Albrecht put his in the back of his Volkswagen. Albrecht used the minicomputer to bring his traveling computing and BASIC show to—he estimated—thousands of students and educators around California. DEC printed and distributed Albrecht's hugely popular book *My Computer Likes Me When I Speak in BASIC*.[103] Albrecht wisely retained the copyright, but DEC benefited because it was promoting the widespread use of its minicomputers through the distribution of Albrecht's book.

My Computer Likes Me, a BASIC primer, sold over 250,000 copies, and from the outset Albrecht declared that his book was foremost about people. The paperback was informal, with a soft cover and pages that featured a variety of hand-drawn cartoons, notes, and phrases alongside teletype printouts. It began, "This book is about people, computers and a programming language called BASIC. We will communicate with a computer, in the BASIC language, about population problems."[104] Albrecht, like Kemeny and Kurtz, put people before computers. He believed that computing should be of the people, by the people, and for the people.

The book's focus on population-growth examples demonstrated Albrecht's conviction that computing was inseparable from other American concerns. Albrecht introduced BASIC computing ideas and commands through a series of connected examples throughout the book, all of which addressed population growth—and overpopulation. This was a fundamental part of American and global dialogue in the early 1970s. In 1970 the Stanford University biologist Paul Ehrlich appeared on Johnny Carson's *The Tonight Show* to warn Americans about the perils of overpopulation.[105] The combination of computing and overpopulation received worldwide attention with the 1972 publication of the best-selling book *The Limits to Growth.* The authors, an MIT-based team of researchers including Dennis and Donella Meadows, had developed a computer simulation of human population and economic growth, production, consumption, and pollution that forecasted global environmental and economic collapse.[106] In fact, the intersections of computing, population, and pollution were powerful themes during this era of the people computing. In addition to the prominence of *My Computer Likes Me* and *The Limits to Growth,* the Huntington Computer Project (to which I'll turn momentarily)—with programs for students and educators written in BASIC and distributed by DEC—developed simulations about pollution and population modeling. A subset of the Bay Area population at the intersection of technology and counterculture cared deeply about computing and the environment. Albrecht helped launch the loose computing education division of the Portola Institute, which employed Stewart Brand, the father of the quintessential countercultural publication the *Whole Earth Catalog.*[107] The *Whole Earth Catalog* was favored by those interested in (among other things) ecology, the environment, and back-to-the-earth communes. Albrecht mirrored these issues in *My Computer Likes Me* when he noted about population that "if the present growth rate persists," there would be "too many" people.[108]

The first edition of *My Computer Likes Me* was snapped up during 1972, the same year that Albrecht expanded his paired BASIC-and-people-computing mission with the publication of the *People's Computer Company.* The *People's Computer Company* was, at the outset, "a newspaper about having fun with computers," and it was accompanied by a People's Computer Center, "a place to do the things the *People's Computer Company* talks about."[109] Albrecht's periodical incorporated photographs, hand-drawn cartoons, teletype printouts, multiple fonts, and handwriting—often all on the same page. It was irreverent and eye-catching and enthralling, all to encourage widespread, interactive

The *People's Computer Company* declared that BASIC was the language of the people computing. Bob Albrecht founded the *People's Computer Company*, and he zealously supported BASIC, together with widespread time-sharing and computing in schools. From the Liza Loop Papers, M1141, Box 3, courtesy of the Department of Special Collections, Stanford University Libraries.

computing. The back cover of the first newsletter featured a cartoon group of diverse individuals: girls and boys, women and men, black and white, with some holding signs that proclaimed, "BASIC IS THE PEOPLE'S LANGUAGE!" and "USE COMPUTERS FOR PEOPLE, NOT AGAINST THEM!"[110]

The *People's Computer Company* aimed to bring computing to the people. The entertaining periodical printed program listings for a variety of games, including the BAGELS number-guessing game; MUGWUMP, which was a game of hide-and-seek on a grid (the player had to find a particular point on a grid); HURKLE, another find-the-imaginary-character game; and INCHWORM, a game for moving an imaginary inchworm around a grid.[111] With the program listings, *People's Computer Company* readers could create, modify, and play these games on their own time-sharing systems. Recall that these games were all played on teletypes, but they were nonetheless absolutely engrossing. A journalist assigned to cover the *People's Computer Company* playfully lamented that, instead of conducting interviews and writing his story, he had "nothing to tell an editor beyong [*sic*] that I spent a total of 28 hours so far just playing games with those seductive machines. This has got to stop. I will, in fact, be back . . . after I can figure out a way to inure myself fromthe [*sic*] seductive call of a clattering teletype. Mugwumps and Hurkles indeed!"[112]

The storefront People's Computer Center facilitated these addictive interactions. The *People's Computer Company* advertised hands-on computing classes at the center, as well as open houses, school visits, and the option to buy computing time. "Buy computer time and do your own thing—play games, learn BASIC, design your own games, zap out math homework. . . . Younger people pay less than older people. From $1 per hour to $2 per hour depending on age and other variables."[113] The *People's Computer Company* also provided additional opportunities to learn about computing by reviewing relevant periodicals, books, films, and museum exhibits about computing, and with a *People's Computer Company* bookstore from which readers could order the program listings or teletype paper tapes for games, as well as books including *My Computer Likes Me*, Kemeny and Kurtz's *BASIC Programming*, and another book on BASIC by Albrecht, his fellow *People's Computer Company* contributor LeRoy Finkel, and Jerald Brown.[114] Contributors to the *People's Computer Company* embraced the potential for producing art by computing, with regular features on BASIC and music, invitations for readers to submit their own teletype art and the program listings that created the art, and the issue titled "The Computer and the Artist," which addressed (among other topics) "how to write poems with a computer," "the computer and the weaver," haiku, interactive computer graphics, and teletype artwork.[115]

The *People's Computer Company* epitomized personal computing before personal computers, and it highlighted the crucial role of DEC, HP, and their minicomputer time-sharing systems in spreading that computing and BASIC during the latter 1960s and early 1970s. Moreover, Albrecht's *People's Computer Company* vigorously endorsed the Huntington Project, whose materials had also been distributed by DEC in its efforts to sell to educational users, and through which thousands of students and amateur computing enthusiasts entered the novel world of computing simulations.[116] In turn, Albrecht used those Huntington Project simulations in his computing roadshow, demonstrating them on—what else—a DEC minicomputer.

DEC, the Huntington Project, and Albrecht Once Again

Networked computing via time-sharing persisted and continued to spread during the 1970s, in large part because it was implemented on minicomputers, especially DECs. Kenneth Olsen and Harlan Anderson, DEC's founders and alumni of MIT's Lincoln Laboratory, initially located a market for their pared-down computers in the early 1960s by focusing on scientists and engineers.[117] Members of this customer group already knew how to program or could learn, so they didn't require the expensive software solutions provided by IBM and its competitors. Furthermore, these scientists and engineers usually did not need the extensive peripheral equipment—keypunches, card readers, printers, disc drives—that drove up the costs of major mainframes. In 1965, DEC introduced the PDP-8 minicomputer; within a few years, the company had sold somewhere between thirty thousand and forty thousand of them (in 1962, there were approximately ten thousand computers in the world).[118] In fact, the PDP-8 family became the best-selling computers in the world from 1973 until the introduction of the Apple II.[119] DEC sold its minicomputers for a fraction of the cost of a mainframe. The PDP-8 was not just miniature in cost but also miniature in size compared with the mainframes. DEC used integrated circuits—now commonly known as "chips"—to build the small, efficient PDP-8s. DEC also produced a time-sharing version of the PDP-8. Thousands of people in schools, universities, research labs, and small businesses gained experience with time-sharing and writing their own programs with the PDP-8 (eventually in BASIC). DEC surpassed the success of the PDP-8 with its PDP-11. The PDP-11 of 1970, which cost about $10,000, matched the computing power of an IBM mainframe of 1965, which cost more than $200,000. The popular PDP-11 could also act as a time-sharing system, and users could write programs in DEC's version of BASIC.[120]

DEC's leadership in minicomputers propelled numerous other companies to enter the market, and, like DEC, these companies offered BASIC on their machines to demonstrate their competitiveness. Between 1965 and 1970, over one hundred new companies or new divisions of existing companies formed to manufacture and market minicomputers.[121] The minicomputer market thrived into the 1980s. Many of these companies created their own versions of BASIC for their mini offerings. Kemeny and Kurtz freely shared BASIC. The 1960s-era Dartmouth BASIC manual, a version of which was commercially published, provided details about the commands and conventions of the programming language.[122] Armed with that information, employees at DEC or HP could create a similar version of BASIC for their own machines. The *People's Computer Company* even compared the implementation of "Dartmouth BASIC" on DEC's EduSystem and HP's minicomputers in its inaugural issue.[123]

Although DEC and HP offered completely different minicomputers, their implementations of BASIC enabled the people using them to easily translate from one system to the other. For the purposes of the *People's Computer Company*, DEC and HP BASIC were nearly interchangeable; for example, when the *People's Computer Company* printed the program listing for the game STARS (a number guessing game), it noted, "STARS was written for [DEC's] EduSystem 20. To run it on an HP 2000, delete line 130."[124] BASIC enabled people to create, modify, share, and store programs on different computing networks; in other words, people produced software using BASIC. BASIC proliferated across time-sharing service centers, time-sharing mainframes, minicomputers, and minicomputer time-sharing systems. Millions of minicomputer users gained familiarity with BASIC, which was especially popular for computing in the educational market. In fact, DEC claimed that over a million kids used its computers in 1972 alone.[125]

DEC received acclaim not just because of its minicomputers but also because it distributed programming for free to anyone interested; the Huntington Computer Project then harnessed DEC's reputation for sharing expertise.[126] The Huntington Computer Project spanned eight years, two universities, hundreds of schools, and thousands of students.[127] Ludwig Braun, a professor of electrical engineering at the Polytechnic Institute of Brooklyn (New York), received a grant from the NSF in 1967 to explore the use of computers

Students used teletype terminals to play the popular computing simulations produced by the Huntington Project. Ludwig Braun headed the project, and its materials were distributed by the Digital Equipment Corporation (DEC) as part of the company's efforts to sell its time-sharing and BASIC-enabled minicomputers to schools. Courtesy of the Computer History Museum.

in the high school curriculum.[128] Braun, like Albrecht, caught the computing bug during the 1960s. He visited Kemeny and Kurtz in Hanover to learn about time-sharing, BASIC, and students teaching the computer, and he, too, became hooked on BASIC as the people's computing language.[129] He organized a project for students around Long Island; some received computing access via stand-alone computers at their schools, while others received time-sharing access. In fact, the Polytechnic may have purchased the GE-265 system from Dartmouth (on which Kemeny and Kurtz first implemented time-sharing) when Dartmouth acquired the GE-625.[130] Over the course of that first three-year project, Braun demonstrated compelling educational value in computing simulations, and the NSF awarded him another two-year grant to develop those simulations.

Braun witnessed firsthand the mesmerizing combination of computing and simulations when he observed students playing POLUT, in which fish were affected by parameters including water temperature, type of pollution, and amount of pollution. Although the students sitting in classrooms were barraged by the noise of loud teletypes, "those kids heard water rippling over the stones in the brook"; when fish began dying, the kids showed "great sadness, and they went back to the drawing board determined to do something to keep the fish alive."[131] By January 1974, the Huntington Project had produced seventeen simulations and support materials for use in primary and secondary school classrooms (see Table 3.4). The programs for the simulations—written in BASIC—were transmitted via paper tape (read by a teletypewriter), and they were accompanied by a teacher's manual, a student's manual, and a resource manual. The simulations reflected the concerns and issues of the day: water pollution, population growth, and domestic policy, for example.[132] Braun and his fellow project leaders planned the simulations to replace laboratory exercises as learning tools in cases where the lab was too difficult, dangerous, expensive, or time-consuming for students to actually learn from it. But the simulations had a side effect: the students who were grouped as lab partners to complete one of the Huntington Project exercises gained computing experience.

After testing and tweaking the simulations during 1970–1972, Braun and assistant director Marian Visich arranged for DEC to publish and distribute the Huntington Project materials in the spring of 1972. In fact, when Braun and Visich secured yet another NSF grant to support the project for 1972–1974, they dedicated a substantial portion of that grant to publicizing and disseminating project materials. They described conferences to "increase teacher awareness" and project staff presentations at local, regional, and national meetings. Braun and Visich had moved from the Polytechnic to the State University of New York (SUNY) at Stony Brook in 1972, and the project moved with them. DEC sold twenty-five thousand manuals during 1972–1973, used by an estimated six hundred teachers and twenty-five thousand students. Braun and Visich anticipated reaching over one hundred thousand students in 1973–1974.[133] By the time that DEC published materials on the simulation BUFLO (about managing bison herds) in 1974, it sold a number of other educational materials alongside it, such as *101 BASIC Computer Games* and *Understanding Mathematics and Logic Using BASIC Computer Games*.[134]

Table 3.4 Huntington Project Computer Simulations

Simulation	Description
CHARGE	Millikan's oil drop experiment
ELECT 1, 2	Simulations of 14 American presidential elections
ELECT 3	Shaping voter attitudes during an election
GENE1	Simple genetics experiment demonstrating Mendelian inheritance
HARDY	Hardy-Weinberg principle of population genetics
LOCKEY	Lock-and-key model of enzyme activity
MALAR	Malaria epidemic
MARKET	Two companies engage in one-product competition
POLICY	Decision making with six political interest groups
POLSYS	Local government responses to community influence
POLUT	Effects of water pollutants and evaluation of antipollution measures
POP	Population modeling
SAP	Simple data analysis package
SCATR	Use of alpha-particle scattering to determine atomic structure
SLITS	Young's double-slit experiment on light waves
STERL	How pesticides and/or sterile male insects affect insect population
TAG	Studying population size on a farm pond

Source: Marian Visich Jr. and Ludwig Braun, "The Use of Computer Simulations in High School Curricula," January 1974, http://eric.ed.gov/?id=ED089740.

Throughout his work with the project, Braun applauded and advocated BASIC.[135] He authored programs by which students immersed themselves in experiments and computing. The materials produced by Dartmouth's Secondary School Project and those generated by the Huntington Project demystified computing for many teachers by providing them with a contained, structured, and clearly documented mode of integrating computing into their classrooms. Furthermore, these materials complemented DEC's and HP's early-1970s push for the educational market; both companies developed systems intended for students and teachers.[136]

Conclusion

During the 1960s and 1970s, access to a computer (whether a mainframe, a mini, or, eventually, a personal computer) was not enough to ensure an enjoyable, productive, or meaningful interactive computing experience. Rather, most people needed a way to transform the computer into a tool for personal or social use. BASIC became the language by which people personalized their computing. They solved homework problems and graded examinations, they composed melodies and poetry, and they immersed themselves in games and simulations. Moreover, BASIC provided a foundation for communities that shared computing enthusiasm and cultivated computing citizenship. BASIC spread like wildfire around Dartmouth's campus and through the woods of New England across the Kiewit Network, catching at K–12 schools, colleges, and universities alike. Bob Albrecht sang BASIC's praises from Minnesota to California and across America through SHAFT, the NCTM, and cool publications including *My Computer Likes Me When I Speak BASIC* and the *People's Computer Company.* In a move that Apple would emulate a decade later, DEC supported its minicomputer market by putting BASIC on its machines and by publishing educational materials, including the clever Huntington Project simulations, that showcased BASIC. In his popular manifesto *Computer Lib: You Can and Must Understand Computers Now,* Ted Nelson urged people about BASIC, "If you have the chance to learn it, by all means do." He recognized that BASIC had been "*contrived* specifically to make programming quicker and easier," yet it was "a very serious language" for "people who want simple systems to do understandable things in direct ways that are meaningful to them, and that don't disrupt their companies or their lives."[137] From its humble origins in small-town Hanover, New Hampshire, BASIC became the language of millions of people computing.

4

The Promise of Computing Utilities and the Proliferation of Networks

THIS CHAPTER REPRESENTS a departure from the rest in this book. The other chapters portray a local, ground-level view of the rise of American computing culture. Chapters 1 and 2 narrated how students taught the computer to create a network at Dartmouth and chronicled the creation of Dartmouth's masculine, creative, games-focused computing. Chapter 3 contended that BASIC (Beginners' All-purpose Symbolic Instruction Code) was the lingua franca of the people computing, and it traced the thread of that language through the northeastern Kiewit Network, across the nation with the symbiotic relationship of the minicomputer marketplace and the Huntington Project, and westward to California via BASIC's peripatetic missionary Bob Albrecht. The following chapters will stay similarly local, to showcase the rise of a statewide public computing network in Minnesota, and finally to examine another richly textured national network built on time-sharing, the PLATO (Programmed Logic for Automatic Teaching Operations) system centered at the University of Illinois.

This chapter, in contrast, provides a bird's-eye view. Here, I want to showcase the national conversations around computing and networks that thrived during the 1960s. America's current Silicon Valley mythology imagines a progression from mainframes to personal computers. IBM dominated computing in the 1960s until the home-brew hobbyists of Silicon Valley (including Steve Jobs, Steve Wozniak, and Bill Gates) offered liberation with their personal computers and software during the latter half of the 1970s. In that story, Americans didn't gain the full promise and potential of personal computing until they could access the Internet during the 1990s. This popular myth nods to the early-1970s genesis of the ARPANET as the origins for American computer networking, and it traces a tidy path from the nascent network sponsored by the Defense Department's Advanced Research Projects Agency (ARPA) to the Internet of today. That mythology is all wrong.

Multiple computing networks proliferated during the 1960s and 1970s. And as soon as people began talking about time-sharing—circa 1960—they began discussing the possibility of a national computing network, comparable to the national telephone network or the national electrical grid. During the 1960s, academics and businesspeople alike grew increasingly impassioned about a national computing utility, or even multiple computing utilities, in which computing services would be delivered across the United States over time-sharing networks. They discussed it at conferences, published articles and books about it, and launched businesses to implement it. They imagined a world in which all Americans benefited from computing access in their homes and workplaces, just as Americans benefited from the comparable national utilities of water, electricity, and telephone.

This chapter recovers this lost era of American computer networking. The first section examines the rise of the computer utility in national discourse during the first half of the 1960s. Here, the role of MIT in Cold War computing research becomes crucial. If IBM dominated the business of American computing during the 1960s, then MIT was America's bellwether for computing research. And two men affiliated with MIT computing, John McCarthy and Martin Greenberger, espoused the idea of a national computing utility. As this section shows, the trajectory of time-sharing at MIT fomented enthusiasm for computing utilities. MIT's choice of GE over IBM as the provider for a significant time-sharing project offered a similar boon to GE in its time-sharing business.

The next section focuses on GE time-sharing as a microcosm of the computing utilities business. The computing utilities industry boomed during the latter half of the 1960s and maintained its momentum through the 1970s and into the 1980s. GE was one of many companies that sold computing as a utility during these decades, and its partnership with Dartmouth College gave it a head start. The computing utilities market did not collapse in the early 1970s, contrary to what some historians have claimed.[1] It is true that GE sold off its computer manufacturing business at that time; however, GE retained its time-sharing service bureau to sell computing as a utility into the 1980s.

The third section dissects the promises and perils of a national computing utility, seen through the eyes of its boosters in the latter half of the 1960s. The MIT professor Greenberger, John Kemeny, the prominent networks thinker Paul Baran, and others elaborated their high hopes for a national network; however, they were cognizant of the technological, legal, economic, and regulatory obstacles to a computing-communications network. This section employs gender as an analytical category to analyze how Cold War academic-industrial masculinity shaped the discourse about a national computing network. National computing network boosters often described its proposed uses in terms of their value to the white middle-class adults of the heteronormative Cold War nuclear family, with a distinction between the associated public and fiduciary responsibilities of the father and the private, domestic, nurturing responsibilities of the mother. In other words, the computing network reflected 1960s middle-class gender roles. I contend that, as a result of this adult gendering of the national network, contemporaneous observers—whether academics, journalists, or businesspeople—overlooked the richly textured personal and social computing practices that students and educators were already creating on their nascent time-sharing networks.

Just as the second section of this chapter documents the many business computing utilities that emerged between 1965 and 1975, the fourth and final section illuminates the many academic computing networks that originated in that ten-year span. These include time-sharing networks at places such as Dartmouth, Carnegie Mellon, and MIT, but other computing partnerships developed in places such as North Carolina, Michigan, and Oregon. In fact, by the early 1970s, there were so many computing networks (and ARPANET was just one of many) that observers began defining and differentiating the various kinds of networks. The purpose of this section is not to suggest that personal and social computing flourished across all these different kinds of networks. On the contrary, the Dartmouth and Kiewit networks and (as we'll see in Chapter 5) the statewide Minnesota network were distinct precisely because they encouraged computing among all community members. In contrast, many of the academic computing networks reserved or prioritized computing resources for those working in the sciences and engineering. Rather, the purpose of this section is to call attention to the proliferation of computing networks from 1965 to 1975. These networks collectively embodied the desire for computer resource sharing, for a community of interested individuals joined by computing networks, in short, a desire for communal computing. That same impetus propelled the push for a national computing utility, the promise of a nation of computing citizens.

MIT and the Computing Utility: Pitches by McCarthy and Greenberger

In 1961, when he delivered his first public lecture on time-sharing, the influential computer thinker and tinkerer John McCarthy declared that time-sharing would make a nationwide computing utility possible. The occasion was the centennial celebration for MIT, which had dominated American computing development during the 1950s. McCarthy explicitly likened a computer utility to the telephone or electricity, a public utility for American citizens.[2]

Three characteristics of McCarthy's 1961 centennial lecture became significant and shaped the public discourse about computing for the next decade: it happened at MIT (which was a bellwether of American computing), it endorsed time-sharing, and it offered the possibility of computing as a public good with the utility model. Indeed, for American technological and business leaders and journalists—and by extension, the American public—MIT, time-sharing, and the computer utility were intertwined and inseparable during the 1960s. When enthusiasts hoped and planned for a computer utility, they looked to MIT as the beacon of progress and possibility in time-sharing and computing networks.

By the time McCarthy delivered his 1961 MIT lecture, the university had already earned a reputation for computing prowess because of two projects: Whirlwind and SAGE (Semi-Automatic Ground Environment). Whirlwind originated during World War II as a project to train pilots to fly new types of airplanes. These aircraft simulators modeled cockpits and aircraft controls with appropriate feedback, and they effectively trained pilots at a much lower cost than flying actual planes. But in 1943, in the heat of World War II, a different simulator had to be built for each new kind of plane. The U.S. Bureau of Aeronautics Special Devices Division invested in a universal flight trainer, and MIT received the initial contract for the feasibility study.[3]

Whirlwind ultimately careened dramatically over budget and time estimates, costing $8 million over eight years rather than the projected $200,000 over two years. But the Whirlwind Project fostered essential research that improved computing speed and reliability, and it birthed the first operational core memory system. Magnetic core memory offered far more reliable and consistent computer memory than the existing vacuum-tube or mercury-line systems. Whirlwind also enabled real-time computing, meaning that the computer responded to external messages as it received them, in seconds or faster. MIT transferred much of the Whirlwind technology to IBM in the early 1950s, a boon to the university in multimillion dollar royalties, and a boon to IBM's growing domination of the computer industry.

The $8 billion SAGE Project created a home for the Whirlwind computers, and, like Whirlwind, it was an expensive project whose long-term value had little to do with its intended purpose. SAGE was planned as a comprehensive, computer-based, U.S. air-defense system with aircraft, missiles, and artillery to protect the nation from the Soviet Union. The momentum for SAGE grew in 1949 when the United States learned that the Soviet Union had detonated a nuclear bomb and that its planes could drop such bombs on the United States.

By the time MIT fully deployed SAGE in 1963, intercontinental ballistic missiles had superseded bomb planes as the urgent military threat, but this massive Cold War project had further stimulated MIT (and IBM) computing. In 1951 MIT's new Project Lincoln, formed for SAGE research and development, absorbed Whirlwind. What became Lincoln Laboratory invested in civilian computing by contracting with IBM, Bell Labs, and many other manufacturers for SAGE. SAGE supported the development of graphical displays, core memories, and printed circuits. MIT and IBM gained computing prominence.

MIT then marked the first one hundred years of its existence in 1961 with the lecture series "Management and the Computer of the Future," befitting its place at the pinnacle of American computing research. The School of Industrial Management (now the MIT Sloane School) sponsored the series, and Martin Greenberger planned it and later edited the lectures and responses for publication as the book *Computers and the World of the Future.* By 1961 Greenberger had worked with computers for nearly a decade; he was an expert at a time when the computer industry burgeoned and transformed. IBM employed Greenberger before he became a professor at MIT, and as an IBM employee, Greenberger helped establish the MIT Computation Center, the centerpiece of which was an IBM computer. During the 1960s Greenberger deployed his industry experience and computing expertise as a prominent public intellectual.

The program for MIT's lecture series presented a who's who of the American scientific and computing elite. Vannevar Bush directed the Office of Scientific Research and Development during World War II; he had invested in American science at an unprecedented rate to support the war effort. Jay Forrester directed Project Whirlwind and then Lincoln Lab, where he led the design of SAGE. Grace Hopper worked with Harvard's computer during World War II before introducing new programming techniques and the computer compiler to the industry during the 1950s. Kemeny spoke, as did the physicist Sir Charles Percy (C. P.) Snow, the author of *The Two Cultures and the Scientific Revolution.* Other speakers included John Mauchly, well known to this community as the cocreator of World War II's ENIAC (Electronic Numerical Integrator and Computer), America's first electronic computer; Marvin Minsky, the codirector of MIT's Artificial Intelligence group; and Norbert Wiener, a leading proponent of cybernetics, an influential approach to scientific problems of control and communication.

McCarthy was in good company, then, when he delivered his centennial lecture "Time-Sharing Computer Systems"; what has been overlooked about McCarthy's lecture is that he advocated for both time-sharing and a public computing utility. McCarthy, just a year younger than Kemeny, had also earned his doctorate in mathematics at Princeton. During the 1950s McCarthy taught at Stanford and then at Dartmouth. In fact, MIT's Computation Center, with its dedicated IBM computer, lured McCarthy away from Hanover in 1958. For McCarthy, time-sharing and the public good were intertwined. Most scholars and journalists point to the emergence of the public utility concept circa 1964, heralded by Greenberger's article about it in the *Atlantic Monthly*. But McCarthy pitched time-sharing for the public in 1961. Why does this matter? People were thinking about time-sharing as a way to bring computing to the people from its earliest days. There was a push for accessibility and use—one that was taken seriously by Kemeny, Kurtz, McCarthy, Greenberger, and their colleagues.

McCarthy defined his proposed time-sharing system, from the outset of his lecture, in terms of users. He declared in his second sentence, "By a time-sharing computer system I shall mean one that interacts with many simultaneous users through a number of remote consoles. Such a system will look to each user like a large private computer."[4] His user focus extended to a specific and significant analogy that foreshadowed his public utility proposal. He explained, "When a user wants service . . . the computer is always ready to pay attention . . . just as the telephone system is always ready for you to lift the receiver off the hook."[5] McCarthy thereby signaled to his contemporaries that time-sharing computing promised the convenience, accessibility, speed, and ubiquity of the international telephone network. He cemented this comparison when he explained that the IBM 7090 computer could handle three thousand consoles, with the implication of three thousand users, simultaneously. At a time when there were only about six thousand computers total in the

United States, the proposal for a computing network with three thousand simultaneous users was striking.[6] Moreover, McCarthy outlined the possibility of a system featuring not just the text-based interface of a typewriter but also the graphical interface of a "cathode-ray tube unit on which the computer can display pictures and text."[7] McCarthy likely had been inspired by MIT's experience building SAGE with graphical displays.

McCarthy underscored the relationship between time-sharing and the public good in his conclusion, a significant point in any lecture to raise the wider implications of the subject at hand. McCarthy expounded, "If computers of the kind I have advocated become the computers of the future, then computation may someday be organized as a public utility, just as the telephone system is a public utility."[8] With his analogy to the telephone system, McCarthy made clear that this computing utility could be supported by local, state, and federal governments for their constituents, as a community necessity comparable to telephone service, electricity, or water. McCarthy also pitched the computing utility in business terms, "as the basis for a new and important industry."[9] Perhaps McCarthy framed time-sharing this way because he was speaking at a lecture series organized by the School of Industrial Management, or perhaps he believed that a computing utility industry regulated by the government would work well. Either way, his focus on "subscribers pay[ing] only for . . . capacity" represented a very different model from the library system envisioned and implemented at Dartmouth, a system that focused on free-to-users service subsidized by the institution.[10]

McCarthy's speech served as the touchstone for the technocratic vision of computing for the next decade: it centered on MIT, it endorsed time-sharing, and it tied a computing utility to business. Indeed, during 1961 a team at MIT led by Robert Fano and Fernando Corbató led the effort to build MIT's Compatible Time-Sharing System (CTSS), which they widely publicized and soon planned to expand.[11] Although Corbató envisioned MIT's system as a resource for the entire university and local Cambridge community, it nonetheless remained the domain of scientists and engineers.

When J. C. R. Licklider, another MIT Centennial participant, accepted the directorship of the Information Processing Techniques

Office at the Advanced Research Projects Agency (ARPA) in 1962, he focused his attention on familiar faces and places, including MIT. The Soviet launch of the *Sputnik* satellite in 1957 jolted the American political, intelligence, and scientific communities. Federal funding poured into science research and science education.[12] Scientific and technological leadership marked another front in the global Cold War. President Dwight Eisenhower established ARPA in 1958 to conduct defense research that would yield results in the long term. ARPA projects thus enjoyed the luxury of time during an era in which many military-funded projects were carefully managed with tight deadlines. One such long-term ARPA project commenced in 1962, a $7 million investment to explore computing for national defense. The psychologist Licklider, who authorized the project, had established the psychology department at MIT and contributed to SAGE.

Licklider's studies of behavior and mind shaped his computing research and his ARPA leadership. He aimed to create computing that would complement people's mental powers and processes. In 1960 Licklider published an essay, "Man-Computer Symbiosis," that elaborated these ideas.[13] His goal of useful, interactive computing (rather than artificial computing intelligence) informed his investments through ARPA. He awarded projects to colleagues who supported his vision and who were based at already-prominent research institutions, including Stanford, Carnegie Mellon, and the University of Utah.

Licklider deemed time-sharing essential to his studies of human-computer interaction, and he deemed MIT essential to time-sharing. MIT's early lead in time-sharing research, combined with Licklider's long-standing and fruitful relationship with MIT, informed his decision to award $3 million of ARPA funds to MIT to build a state-of-the-art time-sharing system. Its acronym stood for either "multiple-access computer, machine-aided cognition, or man and computer," but in short, Project MAC spelled success for MIT and Licklider.[14] Project MAC's popularity provided a compelling proof of concept for a computing utility, and the public discourse around computing utilities took off during MAC's heyday in 1964 and 1965.

Greenberger drew the public's attention to computing as a utility in his May 1964 essay for the *Atlantic Monthly*, "The Computers of

Tomorrow."[15] Greenberger had been ideally positioned to write this piece because he organized the MIT Centennial Lectures in 1961 and then edited them for hardcover publication in 1962 and paperback publication in 1964. Indeed, the paperback run of Greenberger's *Computers and the World of the Future* corroborated the popularity of the topic. In his *Atlantic* article, Greenberger invoked Vannevar Bush's 1945 predictions—in the *Atlantic Monthly*—for the "advanced arithmetical machines of the future."[16] Greenberger celebrated "the remarkable clarity of Dr. Bush's vision."[17] Greenberger thus associated himself with Bush's prominence and brilliance, and he subtly foreshadowed that his own 1964 *Atlantic* predictions would play out in the following twenty years.

What did Greenberger anticipate? He contended, "Computing services and establishments will begin to spread throughout every sector of American life, reaching into homes, offices, classrooms, laboratories, factories, and businesses of all kinds." The first two subheadings of Greenberger's essay delineated his vision: "Analogy with Electricity" and "The Information Utility." Greenberger qualified his analogy to electricity with three cautions: first, computers seemed remote and even ominous to most users; second, computers currently lacked the variety of appliances that had been developed for electrical use; and third, electricity was homogeneous, whereas computing was "dynamic" and "guided by the action of the user." Nonetheless, Greenberger expressed confidence in "the dream of large utilities built around the service of computing systems." Greenberger diplomatically played to both sides of the developing computing utility vision: those who supported a public utility and those who supported private ownership and operation. He acknowledged that "the public-utility format may still prove to be the best answer" because of the substantial capital investment in equipment and programming. He then conceded that "the stimulating effect of free enterprise and competition on imagination and hard work" offered compelling reasons for the private-company format.[18]

Greenberger outlined the benefits of an information utility with uses that reflected his business background and appealed to the *Atlantic*'s affluent, white, male, professional readership: more convenient banking, lower insurance premiums, computerized stock markets, and information systems for hospitals, among others. In short, Greenberger's blueprints for an information utility projected his own experience at MIT and its business school, where Project MAC offered data processing and information sharing—but not communication—for a white, masculine, professional class. Indeed, Greenberger concluded, "By 2000 AD man should have a much better comprehension of himself and his system . . . because he will have learned to use imaginatively the most powerful amplifier of intelligence yet devised."[19]

The year 1965 marked a turning point for transforming the vision of a community computing utility into reality for two reasons: ARPA awarded MIT a contract to expand Project MAC, and American capitalists increasingly viewed computing utility businesses as viable and attractive opportunities. ARPA invested millions of dollars in the new MIT time-sharing system, known as Multics (Multiplexed Information and Computing Service). The Multics team anticipated a system with one thousand terminals, of which three hundred could be used simultaneously. Fano and Corbató publicized the developing system in a 1966 article for the wide readership of *Scientific American*, "Time-Sharing on Computers."[20] They described the "dialogue between user and machine" and "communication among users."[21] The opening page of the article showcased thirty photos of people—mostly men—using teletypes, and the authors explained that those individuals used teletypes for computing in offices and laboratories, "in large 'pool' rooms," and "a few in private homes."[22] Fano and Corbató notably explained the relationship between community and computing utility as a symbiosis. They concluded, "The coupling between such a utility and the community it serves is so strong that the community is actually part of the system itself."[23] They acknowledged that "because such a system binds the members of a community more closely together, many of the problems will be ethical ones," such as questions of access, regulation, and safeguards against misuse; however, they left those questions unanswered.[24] In the introduction to the same issue of *Scientific American*, McCarthy assured readers that computers, "far from robbing man of his individuality," would instead "increase human freedom of action."[25] He, too, predicted "computer consoles in every home and connected to public-utility computers through the telephone system."[26]

The academic community's enthusiasm for a computing utility was perhaps exceeded by that of American capitalists, who created a computing utility bubble in the latter half of the 1960s. Starting in 1965, an abundance of articles and books touting computing utilities appeared.[27] *Fortune* magazine proclaimed time-sharing and the utility the "hottest new talk of the trade" in 1965.[28] By 1967, business analysts valued the computer utility market at $15–$20 million per year, and companies offering time-sharing computing services appeared in cities across North America and Europe. University Computing Company (UCC), based in Dallas, Texas, epitomized the computing utility bubble; from 1967 through 1968 its market valuation leaped from $1.50 to $155 per share. The computing utility business lured established and new companies alike. Newcomers such as UCC, Tymshare (San Francisco), Keydata (Boston), and Comshare (Ann Arbor, Michigan) joined behemoth IBM and the newly prominent GE in the competition for information utility business.

GE and the Business of Computing Utilities

Individuals at Dartmouth and GE worked together from the beginning of the Dartmouth time-sharing endeavor. GE had not envisioned something like time-sharing for its line of 235 or Datanet-30 computers.[29] At the time, GE's computer business, based in Phoenix, Arizona, focused on making computers for particular clients and purposes. For example, GE had entered the computer industry only a few years earlier, in 1956, when the company received the contract to build an Electronic Recording Machine Accounting system, widely known as ERMA, for the Bank of America. This system enabled local bank branches to process checks digitally with the creation of a computer-readable font. Despite the success of ERMA, GE leadership discouraged expansion of the computer division, so the Phoenix group produced machines that originated in particular client requests.[30]

The Dartmouth delegation that visited GE Phoenix when Kemeny and Kurtz were evaluating computer vendors brought the concept of time-sharing to the company's attention.[31] According to Homer Oldfield, who initially managed GE's Computer Department in Phoenix, the GE executives at the company's Lexington Avenue headquarters in New York City frowned upon the computer contract with Dartmouth, but the work nonetheless proceeded.[32] Oldfield chronicled the day that GE employees Bill Bridge and Ray Barclay finalized the GE-225 and Datanet-30 for shipment from Phoenix to Hanover. Bridge predicted, "This may turn out to be the most important shipment yet made by the Computer Department."[33] Regardless of whether Bridge made that exact statement, Oldfield's choice to emphasize that day in GE history demonstrated his recognition of the importance of Dartmouth and time-sharing to GE during the 1960s and 1970s.

During 1964, as Kurtz and his Dartmouth students successfully implemented time-sharing on campus, GE and Dartmouth remained in close contact. GE delivered its computers to Hanover in February, and the Dartmouth team realized time-sharing success in May. Then GE delivered a 235 computer to campus in September to replace the 225 machine; the 235 increased the response speed on the time-sharing system by a factor of five.[34] Kurtz, Kemeny, and their students remained in contact with GE programmers throughout the year, seeking answers to their questions about programming the GE machines.[35]

Dartmouth and GE publicized their time-sharing success at the Fall Joint Computer Conference (FJCC) in San Francisco in October 1964. Dartmouth organized a booth at the conference that housed three teletypes connected via long-distance telephone line to the GE computers in Hanover. GE paid all costs associated with Dartmouth's booth at the FJCC—including the expensive long-distance telephone costs—except for the cost of two Dartmouth students to travel to San Francisco to manage the booth.[36] GE recognized that this demonstration provided excellent publicity for its computer business and for the new mode of computing known as time-sharing. Kurtz updated Kemeny, who was enjoying academic leave in Israel, about the FJCC plans and results throughout the autumn of 1964.[37]

Kurtz conveyed his satisfaction at the results of the FJCC in one of these missives to Kemeny. He noted that the students, Ken Lochner and Bill Zani, handed out several hundred fliers and BASIC manuals. The four-page flier described Dartmouth time-sharing to conference attendees. More importantly, nearly one thousand conference attendees either observed the teletypes in use or experienced time-sharing themselves at one of the teletypes. Kurtz considered the FJCC a "smashing success."[38]

Kurtz also reported to Kemeny another significant GE computing feat of the fall: MIT had selected GE—instead of IBM—as the computer provider for the time-sharing component of its major computing effort, Project MAC.[39] During October, Kurtz journeyed down to Cambridge (yet again) to deliver an academic talk on Dartmouth time-sharing to his computing colleagues at MIT. He noted that Fernando Corbató expressed kudos on the Dartmouth group's accomplishment, and he mentioned "the big news . . . that project MAC is going GE-635."[40] Kurtz's coverage of the GE-MIT news mirrored the buzz that MIT's choice had been generating.

The contemporaneous media attention and subsequent academic attention to the GE-MIT connection obscured the fact that GE developed a very successful computing utility business based on Dartmouth's time-sharing system. During their post-fall-quarter break in 1964 (spanning from December 1964 into January 1965), two Dartmouth students spent several weeks at the GE computer department in Phoenix. The Dartmouth and GE accounts differ on the purpose of that visit, but it is clear that at the end of it, GE had a working implementation of the Dartmouth Time-Sharing System (DTSS) on its computers in Phoenix, along with documentation for the time-sharing system, and BASIC. Dartmouth folks thought the students were in Phoenix to install a working version of the DTSS at GE.[41] GE folks contended that their engineers invited the Dartmouth students to help improve the Dartmouth system's reliability.[42] Don Knight, a GE engineer, recalled that the GE programmers and Dartmouth students reviewed the code line by line. The Dartmouth students explained what they were trying to accomplish, and the GE programmers instructed, "Now here's how you should have done it."[43] In the process, GE obtained a reliable version of Dartmouth time-sharing.

After the Dartmouth students facilitated the implementation of Dartmouth time-sharing at GE Phoenix, GE wasted no time in developing and promoting its time-sharing business. In early 1965, GE launched its time-sharing business by selling and leasing GE time-sharing computers, and by opening time-sharing service centers through which individuals, small businesses, and educational institutions could obtain computing.[44] GE soon counted clients ranging from banks and aerospace companies to automobile manufacturers.[45]

GE marketed the immediacy, flexibility, and utility of time-sharing to its customers. A 1968 brochure for a GE-400 time-sharing system prominently featured a suit-and-tie-clad businessman focused on his teletypewriter. In one image, he is standing with his left foot perched jauntily on a chair while reviewing a printout. The image conveyed a mix of professionalism and casual confidence: although he wore a tie, this businessman easily mastered computing.[46] The brochure identified that time-sharing was essentially "computer-sharing," and the brochure copy sold potential clients on GE time-sharing and time-sharing systems as fast, easy to learn, and productivity-enhancing. GE emphasized the "conversational" nature of time-sharing, by which a user inputting a program would receive a response on the teletypewriter in seconds. Altogether, GE sold the portability, ease, and responsiveness of time-sharing by proclaiming "hundreds of your people can virtually *put a computer in their laps.*"[47] Indeed, that phrase captured the sense of personal productivity and interactive computing that time-sharing enabled.

GE promoted BASIC with separate materials. Its strategy of highlighting BASIC apart from time-sharing targeted different audiences. While executives might have considered a GE time-sharing system for their corporations, small-business owners, scientists, engineers, and academics may have sought out BASIC—and therefore a GE time-sharing service—for their work. The brochure for GE BASIC celebrated the ease of language acquisition ("BASIC is for beginners") and the computing power held within the deceptively simply structure ("BASIC is for experienced programmers, too"). The back of the brochure featured user endorsements: "I do not program by profession, but use BASIC in the solution of everyday problems when necessary." "Our reduction in time with the computer is about 20 to 1." And my personal favorite for its humorous take on providing an endorsement, "Excellent; greatest thing since matches!"[48] The other computing languages available in 1965, such as FORTRAN and ALGOL, were challenging to learn. By advertising the immediate feedback from learning and using BASIC, GE also showcased the responsiveness and convenience of time-sharing.

GE's sales and marketing efforts proved effective because by 1968, GE featured time-sharing centers in twenty-five locations across the United States and around the world; GE counted over fifty thousand individual users of its time-sharing service centers in the United States.[49] Individuals could purchase time-sharing computing in Cleveland, Los Angeles, Boston, and beyond.[50] Students from the London School of Economics, Imperial College London, and Glasgow's University of Strathclyde accessed the GE center in London after it opened in August 1967.[51] Centers in Paris, Sydney, and Milan followed in 1968. By 1970, over one hundred thousand customers used GE time-sharing in Europe.[52] These time-sharing service centers, with BASIC on GE computers, offered the possibility of personal computing to thousands of people.

Alexander (Alex) Conn, a member of Dartmouth's class of 1968, embodied the connections between the college and GE. He wandered into the College Hall teletype room during the fall of 1964, "sat down, played around a bit, asked some questions, and soon was programming in BASIC."[53] He used DTSS throughout his undergraduate years as an engineering major and for his master's thesis at Dartmouth's Thayer School of Engineering. During the summer of 1968, Conn worked for Bull-General Electric in Paris. GE had just opened its Paris time-sharing service center in February. Conn's responsibilities included "helping people learn BASIC on what was the commercial version of DTSS that GE was selling."[54] He eschewed the various Dartmouth and GE manuals, and he avoided the recipe analogy. Instead, he opted to explain computer programming generally and BASIC in particular through flowcharts. He urged his students to break down their business, science, or engineering problems into their most basic component steps, diagram the steps and relationships among them, and then translate those diagrams into programs.[55] Conn also cowrote an article about the state of computing in the American educational system for the French publication *Arts et Manufactures.* Conn recalled that he wrote the article in English, and his manager at Bull-GE translated the article to French. Conn subtly showcased his Dartmouth loyalties. The article neglected to mention GE, but the section on time-sharing highlighted the college.[56]

The Dartmouth-GE relationship persisted throughout the 1960s, including a joint Dartmouth-GE project to expand Dartmouth's time-sharing system. Although the terms of the collaboration changed between 1966 and 1969, Dartmouth ended up with an expanded and improved time-sharing system on a GE-635 computer in the Kiewit Center.[57] GE obtained the expertise to deploy that time-sharing system, both at its time-sharing service centers and for clients purchasing GE time-sharing computers. GE also effectively gained the Kiewit Center as a time-sharing service center and sold time on the Dartmouth system to clients around New England.[58] From 1966 through 1968, Dartmouth and GE programmers worked in concert to develop the new system. A 1966 project memo assigned various parts of the project to Dartmouth programmers and to GE programmers in Phoenix, Hanover, and Falls Church, Virginia.[59] The ongoing communication among the Dartmouth and GE groups ensured that GE learned firsthand about the struggles and solutions for building and maintaining a more powerful time-sharing system.

GE trumpeted its Dartmouth connection to the world via color advertisements and reports on computing in education. A full-page color magazine ad from 1968 featured a large picture of John Kemeny and eight Dartmouth students in Dartmouth attire casually gathered around three teletypes. The ad boldly proclaimed, "The men of Dartmouth have an incredible advantage over the mathematicians who worked at Los Alamos in 1945. Ask Professor John G. Kemeny. He was there."[60] This ad invoked the premier scientific and technological achievement of a generation: the construction of atomic bombs at Los Alamos as part of the Manhattan Project; and then the ad suggested that Dartmouth students somehow enjoyed an advantage over those brilliant physicists—an intriguing premise. The ad explained that "Dartmouth and General Electric . . . [took] the mystery out of laymen using the computer."[61] GE showcased time-sharing, BASIC, and the success of computing at Dartmouth. GE also sought to bolster its reputation in the time-sharing arena through a series of 1968 reports titled "Computer Time-Sharing on Campus: New Learning Power for Students."[62]

The GE reports on educational computing publicized the findings of the President's Scientific Advisory Committee on computers in higher education, widely known as the Pierce Panel. John R. Pierce, a prominent researcher at Bell Labs, chaired this panel, and both Kemeny and Kurtz served on the panel. The panel issued its recommendations in a February 1967 report that commanded attention across the education and business worlds.[63] GE emphasized the Pierce Panel's findings, which urged that computing was possible and necessary for all college students. GE reprinted the Pierce Panel's declaration that "undergraduate college education without adequate computing is deficient education, just as undergraduate education without adequate library facilities would be deficient education."[64] The comparison of computing to library access surely came from Kemeny and Kurtz, who had used the same analogy when proposing time-sharing to the Dartmouth trustees. The GE report cited the activities at Dartmouth College, Dartmouth's Amos Tuck School of Business, and the Dartmouth Medical School as three of the seven examples of outstanding uses of computers in higher education. The GE report also emphasized the power of BASIC as a computing language, and the "payoff of Dartmouth's computer investment."[65] At a time when competitors including Digital Equipment Corporation (DEC) and Hewlett-Packard (HP) were flooding the educational market with their time-sharing minicomputers, GE aimed to maintain its stronghold in the time-sharing realm.

Ultimately the business market and the academic vision for computer utilities were prominently linked; through Multics, the fortunes of MIT, IBM, GE, and Bell Labs became entangled. During the 1950s and the first half of the 1960s, MIT and IBM enjoyed a close and mutually beneficial relationship. IBM licensed—and paid handsomely for—much of the computing technology that MIT's Lincoln Lab produced for Whirlwind and then SAGE. A state-of-the-art IBM mainframe anchored MIT's Computation Center, and the CTSS expanded by Project MAC ran on IBM machines. But while the academic community became increasingly enamored of time-sharing during the 1960s, IBM was investing heavily in its System / 360 project as a massive and dramatically overhauled computing system. System / 360 guzzled all available IBM resources, and the company hesitated to respond to MIT's growing time-sharing needs. Meanwhile, because of its relationship with Dartmouth, GE appeared more responsive to academia's time-sharing dreams. MIT opted to build Multics with GE computers. MIT's choice of GE over IBM for its Multics machines stunned the business community, and MIT's partnership with Bell Labs for Multics programming amplified the symbolic significance of the project.

MIT, GE, and Bell Labs became the polestars for the promise of time-sharing and computing utilities; at the time and since, the ways in which journalists and historians alike focused on the struggles of MIT's prominent Multics project masked the growth of the computing utilities business and GE's success in that realm. Although MIT decision makers perceived GE as agile and adept in developing time-sharing, GE's expertise initially heavily depended on that of Dartmouth students. Moreover, the Dartmouth system served only about thirty simultaneous users at the outset. Thus, GE employees struggled to create the computing infrastructure to meet the Multics goal of hundreds of simultaneous users. In contrast, while MIT and GE grappled with Multics, Dartmouth students and GE employees successfully collaborated to expand the college's time-sharing network. The Bell Labs employees tasked with programming Multics were, like their GE colleagues on the MIT project, overwhelmed. Bell withdrew from Multics in 1969, and a year later GE withdrew from the mainframe business. IBM had belatedly jumped on the time-sharing bandwagon and then lost an estimated $50 million on the venture. These struggles led journalists at the time (and scholars since) to mark the "demise of the time-sharing industry" around 1970.[66]

Yet time-sharing networks and the business of computing utilities continued to grow and thrive during the 1970s. GE's sale of its mainframe computer division to Honeywell in 1970 received significant attention. What went quietly unobserved, however, was that GE retained its successful multimillion-dollar-per-year time-sharing business when it sold its computer department.[67] Since then, that GE even had a thriving computing utility business has largely been forgotten, despite the fact that it continued to be profitable into the 1990s.[68] One notable exception: in his widely popular 1974 work on computing for the people, *Computer Lib / Dream Machines*, Ted Nelson tipped his hat to GE, acknowledging its provision of interactive computing centers across the United States and Europe.[69] What had begun a decade earlier as an experiment in a small liberal arts college had grown into a multimillion dollar business for GE. Tymshare also persisted in offering its computing utility services throughout the 1970s; by 1977 Tymshare's network TYMNET featured over two thousand nodes that served one thousand simultaneous users.[70] In fact, revenues from the time-sharing industry steadily increased every year until 1983, when these networked computing services were challenged by a growing personal computer market.[71]

The Promises and Perils of a National Computing Network

During Dartmouth College's dedication of its new Kiewit Computation Center in December 1966, Kemeny spoke confidently about the future benefits of a national computing network, and the *New York Times* reported it—in starkly gendered terms. Kemeny sketched an America in which every household featured a computing terminal. Children would complete their homework by computer; in fact, Kemeny's daughter already received home computing time as a treat if she was a "good girl."[72] The reporter did not elaborate on what constituted "good" behavior for a girl; however, for middle-class Cold War Americans, the phrase likely connoted a child who was quiet, docile, obedient, polite, and well mannered. Thus, Kemeny's casual reference to his daughter reinforced the notion that computing could reward traditional feminine gender stereotypes. Kemeny further elaborated these gendered norms for the women in the home computing family. The *New York Times* journalist described Kemeny's vision that "housewives would use the machine to program all their chores most efficiently, prepare dietetically balanced menus, check prices for a particular item at all stores in their neighborhoods, place orders, do their banking, order specific home television shows, and for diversion, attain an advanced degree at an university, all via computer and without leaving the home."[73] For all the future possibilities of computing, Kemeny—a male professor at an all-male college—imagined that a national home computing network would merely simplify and streamline the work of housewives. In Kemeny's view, which mirrored the wider American cultural norms around the nuclear family, women would still be responsible for household chores and for the work of obtaining and preparing food—always with an eye toward maintaining the economical household budget. Perhaps the stretch here was the notion that women would obtain advanced degrees "for diversion." Kemeny acknowledged women's need for intellectual stimulation; however, he underscored that housewives belonged in the private, domestic sphere because their advanced academic work would be completed "without leaving the home." Kemeny's discussion of domestic computing made explicit what had been implied in other conversations about a national computing utility.

Most commentators focused on the business uses of such a utility, and the benefits for men in their roles as public professionals and nuclear heads of household. In his 1964 *Atlantic Monthly* article, Greenberger focused on benefits to "user industries," such as banking, retail, and insurance. He speculated on "computer-managed markets," as well as the promises of simulation and modeling.[74] Although Greenberger employed the neutral language of "customers," he revealed his own—and the broader societal—gendered assumptions that associated masculinity with work outside the home, and with financial responsibility for a family. Greenberger elaborated, "As the customer's family expands, as his children approach college age, as they become self-supporting, as he approaches retirement, and so on, his insurance requirements change."[75] He delineated the benefits of an information utility in terms of its benefits to men as professionals and as providers for their families of dependents.

Greenberger underscored his gendered vision for a computing utility when he briefly enumerated its additional promises and possibilities: "medical-information systems for hospitals and clinics, centralized traffic control for cities and highways, catalogue shopping from a convenience terminal at home, automatic libraries linked to home and office, integrated management-control systems for companies and factories, teaching consoles in the classroom, research consoles in the laboratory, design consoles in the engineering firm, editing consoles in the publishing office, computerized communities."[76] What stands out in this list is that Greenberger suggested only one exclusively domestic use for computing: shopping from a "convenience terminal" at home. Shopping was (and continues to be) stereotypically a feminine domain, so Greenberger's association of shopping with computing bolstered gender norms, and a gendered divide. Greenberger amplified this divide with his use of the adjective "convenience" to describe the at-home terminal. Convenience connoted the domestic appliances of the day, machines that supposedly eased the labor and responsibilities of the wife and mother.[77] In fact, other historians have emphasized how the postwar middle-class emphasis on suburbia and its consumption of homes, appliances, cars, and clothing all reinforced the heteronormativity of the nuclear family.[78] A "convenience terminal" for shopping aligned computing with those norms.

As the decade progressed, other commentators similarly focused on adult men as the primary users of a computing utility. In his 1966 book *The Challenge of the Computer Utility*, Douglas Parkhill, a computing expert for the MITRE Corporation, depicted the future with networked computing. Like Greenberger, Parkhill described the transformation of finance with characteristics including "on-line teller terminals" and an "audio response system."[79] He envisioned a "medical-information utility [that] could remember the case histories of millions of patients and make this data available in a matter of seconds," as well as a law utility "containing all of the rulings, laws, case histories, and procedures that are relevant to any particular need."[80] Parkhill predicted "automatic publishing" with an author "writing at the keyboard of his personal utility console," and he claimed (like many others) that computing would transform education by offering students "an unprecedented degree of individual attention."[81] With the exception of education, all of these endeavors were professions dominated by men.

Parkhill expanded Greenberger's comment about convenience shopping into an entire section on computerized shopping; however, Parkhill's sketch emphasized the masculine characteristics of a computing utility shopper. Parkhill explained, "For those readers who envisage a new huckster's paradise, comparable to the 'wasteland' of American commercial television . . . the author hastens to point out that the consumer at his computer utility console is not likely to be the same passive spectator against whom the typical television commercial is directed."[82] No, Parkhill's computer utility user was "an active searcher after information," and he employed "powerful analytical tools." Perhaps Parkhill's imagined "passive spectator" was a housewife taking a break from her exhausting domestic responsibilities, or one of her children plopped on the floor after a long day at school. Regardless, Parkhill clearly conveyed that his computer utility shopper was active and empowered; this shopper was comfortable calculating retailer markups and prices per ounce. He sought out consumer testing comments and customer satisfaction surveys to make the most informed decision. In other words, the computing utility transformed shopping into a masculine endeavor by enabling active searching, quantitative analysis, thorough investigation, and careful consideration.

In his oft-cited 1967 essay "The Future Computer Utility," Paul Baran fused the promise of domestic computing with the presumed masculinity of the utility's primary users. At first glance, to twenty-first-century readers, Baran's prognostication may seem gender neutral. He described using a home console to shop, send messages, pay bills, prepare Christmas cards, and track birthdays. Certainly, these are all activities that men and women do today with their myriad computing devices. Yet, Baran offered telltale clues about the presumed gendering of the user. He shopped for the "advertised sport shirt," a piece of masculine apparel. He paid bills—and taxes—via the console; paying taxes especially denoted the civic responsibility of the masculine family breadwinner. And finally, Baran played on the trope of the clueless husband forgetting his wedding anniversary, thereby triggering an unwanted wifely reaction: "The computer could, itself, send a message to remind us of an impending anniversary and save us from the disastrous consequences of forgetfulness."

Baran left readers to imagine those "disastrous consequences"—a frustrated wife, a hurt or angry one, a wife who abstained from her purported kitchen or bedroom responsibilities? Whatever they were, a computing utility at home would prevent these consequences, promised Baran.[83]

Although he touched on the promises of the computing revolution, Baran identified "privacy and freedom from tampering" as paramount considerations in the creation of a "national computer public utility system"; his concerns were echoed in the computing utility discourse throughout the latter 1960s.[84] Baran explained, "Highly sensitive personal and important business information will be stored in many of the contemplated systems. . . . At present, nothing more than trust—or, at best, a lack of technical sophistication—stands in the way of a would-be eavesdropper. . . . Hanky-panky by an imaginative computer designer . . . or programmer could have disastrous consequences."[85] Baran urged a "fresh regulatory approach" to the privacy challenges posed by computing utilities, and he cautioned his contemporaries who dismissed this concern "as tomorrow's problem" that "it is already here."[86]

Parkhill painted a more detailed picture of the potential problems resulting from networks that contained "a complete record from birth until death of even the most private affairs of everyone," a record that seemed more likely in light of the October 1966 federal report that recommended the creation of a National Data Center.[87] Indeed, the specter of the National Data Center debate haunted subsequent discussions of privacy and civil liberties in the era of burgeoning computing networks, as did the Federal Communications Commission's (FCC) 1966–1967 formal public inquiry into computers and communications, and contemporaneous debates about patents, copyrights, and the nature of intellectual property in the computing world.[88] In the face of intense public and congressional opposition to the National Data Center, the proposal was set aside; however, the debate lingered for those considering the perils of a national computing network. Parkhill feared blackmail and industrial espionage, but he was most especially sensitive to "the possibilities for political repression inherent in a system in which one could not purchase so much as a stick of chewing gum without immediately revealing one's identity and whereabouts to the central computer."[89] He expressed grave concern about the computer utility as "an instrument of total political control."[90] Parkhill exhibited awareness of the Orwellian nature of his warnings, with references to "an uncomfortable aura of '1984' in all this," and to "big brother."[91] Nonetheless, Parkhill perceived great danger "if we were ever to permit any group to manipulate the contents of the public files for private advantage or to tap into an individual's private files without that individual's permission."[92] He concluded, "The computer utility unfortunately could enormously increase the efficiency of totalitarian control and it is this fact that creates the new danger."[93] Parkhill's expression of these concerns was in no way radical. His considered analysis came at the end of his book about the computer utility, and he was writing at a time when the threat of the Communist Soviet Union, and the memory of Nazi Germany, loomed large (he referred to both in his discussion of social implications). Like Baran, Parkhill urged the federal government to demonstrate "strong clear leadership" in shaping the nascent networks.[94]

Just as he had organized the influential lecture series at MIT in 1961 on the future of computing, Greenberger continued to stake his intellectual claim on computing and communications by organizing another lecture series during the 1969–1970 academic year. By that time, Greenberger had traded his MIT appointment for a professorship at Johns Hopkins University, and he parlayed his proximity to Washington, DC, into an increasing stake in the public policy world. In fact, the distinguished DC think tank the Brookings Institution cosponsored Greenberger's 1969–1970 series, in conjunction with Johns Hopkins. In the preface to his edited collection of the lectures, entitled *Computers, Communication, and the Public Interest*, Greenberger explicitly connected his past and present endeavors. "In the years since 1961, the computer has fulfilled and surpassed many of the expectations expressed in the MIT lectures."[95] He noted that more than a thousand people attended some of the 1969–1970 lectures, which were free and open to the public, held at the Johns Hopkins School of Advanced International Studies. Afterward, invited participants strolled across the street for dinner discussions at Brookings (documented in the edited volume). In addition to Greenberger, three other men participated in both the MIT and Hopkins series: Kemeny as speaker both times, Herbert Simon as speaker both times, and Alan Perlis as a speaker at MIT then discussant at Hopkins. In 1961 Kemeny delivered a lecture titled "A Library for 2000 A.D.," and in 1969 he drew on his DTSS expertise to speak about time-sharing networks.[96] Simon, a professor of computer science and psychology at Carnegie Mellon University, gave a lecture titled "Simulation of Human Thinking" at MIT and one named "Designing Organizations for an Information-Rich World" at Hopkins. Like Simon, Perlis taught computer science at Carnegie Mellon, where he also directed the computer center. In 1961 Perlis lectured on computing and universities, and in 1969–1970 he responded to Kemeny's paper on large time-sharing networks.

Greenberger's and Kemeny's endorsement of computing networks, and their enthusiasm for the value of computing utilities, remained strong throughout the 1960s; yet while Greenberger urged caution at the close of the decade, Kemeny presented an optimistic front, and as a group, the Hopkins lecturers continued to reinforce heteronormative Cold War gender roles in their treatments of computing networks. Greenberger reckoned that computing had failed to live up to its early promises, even the promises of only a decade prior, because people had held it at arm's length. He observed, "We tended to view our precocious computer during its early years with awe and adulation. We treated it as something special—a thing apart—because of its unique abilities and great promise. . . . We would have done better to have prepared it (and ourselves) more adequately for the role it had to play."[97] Greenberger declared the unfolding computing saga to be a matter of public concern and public interest, as had Baran and Parkhill before him.

In his lecture "Large Time-Sharing Networks," Kemeny, who had recently stepped into the presidency at Dartmouth College, forecasted the societal changes engendered by such networks, and he recommended the formation of a national agency to facilitate the creation and effective use of those networks to improve society. The new Dartmouth president characterized large networks as those with over a thousand simultaneous users, and he tendered some ideas about how to technically implement such a network. He speculated on the transformation of the medical profession and high school guidance counseling, the development of videophones to replace in-person meetings and conferences, and the proliferation of business offices to accommodate the convenience of employees in suburbs and small towns. He concluded by calling for a National Computer Development Agency (NCDA) "to develop hardware-software systems useful on a national scale but too complex or otherwise costly to attract commercial interest."[98] According to Kemeny, such an agency would support standardization, thereby facilitating connections among multifarious hardware and software systems on the network, and it would spread development costs among states and cities, so that no one locale bore burdensome expenses.

Kemeny pitched the NCDA as an agency devoted to improving society; he acknowledged that "major computer companies" were understandably "profit motivated" rather than service motivated, and that universities focused on research and teaching.[99] Kemeny argued, "A prestigious agency with sufficient financing and a strong commitment to attack social problems would have an immense impact. . . . It is a rare opportunity for the federal government to take an imaginative, effective, and yet inexpensive step in the right direction."[100] During the discussion, Kemeny owned his reputation as an "incurable optimist" in imagining the creation of this federal agency; however, his fellow panelist and discussant Perlis corroborated, "If we do not follow Kemeny's suggestion, our mishandling of the computer's promise will probably not kill us, but it will give rise to questions in the future as to why intelligent and coordinated action was not carried out by a wealthy society devoted to the advancement and freedom of man."[101] Kemeny, Perlis, and many of the other participants in the Hopkins-Brookings lecture series concurred that more immediate work needed to be done to develop, regulate, and direct the activities of a national computing network.

Kemeny sprinkled gendered assumptions throughout his high-minded and optimistic advocacy of computing for service; although he narrated dramatic societal changes, his references to traditional gender stereotypes grounded his claims and reassured his audience, and they anchored the promise of a digitally networked nation to middle-class Cold War gender roles. Kemeny continued to preach the gospel of computing in the home, just as he had in 1966 upon Kiewit's dedication, and he continued to highlight the hypothetical housewife. "The terminal at home will be used to provide valuable shopping information to housewives, such as which nearby stores carry size 7½ AA ladies' shoes and what they charge for them."[102] Here, rather than invoking the woman shopping for family meals, Kemeny invoked the gendered stereotype of a woman devoting her energy to shoe shopping. He contrasted that frivolous image with the serious "customer who [had] printed out at his terminal news items on topics of particular interest to him."[103] The female user of the home terminal was the housewife, shopping for shoes, while the male user was the customer, the domestic resident in charge of purchasing, spending, and saving, and he sought news, not shoes.[104]

When Kemeny tackled how computing would change commuting, he commented, "Another reason why a man goes to his office is because that is where his secretary is. But . . . before very long dictated letters will be automatically transcribed by computer." During the 1960s, secretarial work was pink-collar; most secretaries were women, and their labor was lower paid and viewed as lower status than male-dominated blue-collar or white-collar work. With his remarks, Kemeny may again have been playing to the audience's implicit understanding that a man went to work to see his secretary because she was young, attractive, and at his beck and call. He certainly declared that her labor was easily replaceable by the computer, while simultaneously implying that the businessman's work would only be eased by the computer, never replaced.

Finally, Kemeny claimed that computing would offer new opportunities for women—while ultimately reinforcing their Cold War gender roles. He believed that nationwide networked computing would offer women "continued education in the home and a choice of meaningful part-time employment either at home or nearby."[105] Although this may have seemed like a change from his 1966 association of domestic computing with wifehood and motherhood, his next sentence betrayed that. Kemeny concluded, "For the first time a woman will be able to keep up with her profession without sacrificing her role as wife or mother."[106] Kemeny conceived of women, particularly affluent women, first as wives and mothers. He failed to acknowledge the nearly 40 percent of American women already in the labor force (full-time or part-time) in 1970. Kemeny imagined his prediction for women as progressive, yet he tethered networked computing to restrictive gender roles that contained women in the narrowly circumscribed Cold War nuclear family.

Kemeny cited examples from Dartmouth computing during his lecture, but he missed the vivid patchwork of practices pursued by all students and educators on the Kiewit Network; his speech mirrored the public discourse around national computing utilities that—with its foundation in adult gendered norms and its focus on the future—overlooked the diverse personal and social computing activities that had emerged during the 1960s. The Dartmouth president described his college's time-sharing system in terms of the happiness of its users and their transformed computing knowledge, and he, yet again, exemplified computing recreation by association with football (analyzed in Chapter 2). His—and Dartmouth's—user focus had shaped time-sharing at the college from the beginning, and continued to inform the Kiewit staff's approach to managing its network. Kemeny boasted that Dartmouth's "well-designed system [kept] customers happy" and "improve[d] significantly the acceptance of computers."[107] Kemeny also bragged about gaming on the Dartmouth network, a source of pride and distinction from other colleges that attempted to bar recreational computing. He speculated, "By 1990, I envision two people sitting at terminals quarterbacking opposite football teams in highly realistic matches simulated by computer and displayed with appropriate visual and sound effects."[108] Surely this possibility piqued the interest of his audience members, yet Kemeny sold short the Kiewit Network. He neglected the art, poetry, and music composed, the banners and letters printed, the tens of other games and hundreds of other programs developed, and the communities that formed to share their problems, solutions, helpful hints, and enthusiasms.

The Proliferation of Academic Computing Networks

If the 1960s marked the decade of enthusiasm for computing utilities, reflected in the rapid growth of businesses (like GE) that offered time-sharing to their customers, then the hallmark of 1965–1975 was a proliferation of computing networks, an abundance observed by contemporaries in the early 1970s. By the early 1970s, commentators thought that multiple time-sharing networks could make up a national computing utility (like multiple local electrical companies), that there would be several national networks (like the major television networks of the 1960s), or that there could be one national provider (like American Telephone and Telegraph [AT&T] for telephone service in the 1960s). However, by 1970 there was indubitably discussion of national networks, and Greenberger himself noted, "The practice of calling time-sharing networks 'information (or computer) *utilities*,' a term that Western Union and a number of new companies have adopted in their advertising and shareholder reports, has reinforced the FCC's interest [in regulation]."[109]

In his 1960 MIT Centennial Lecture, McCarthy had proposed the tandem ideas of time-sharing and computing utilities; just five years later, academic leaders across the United States formed an organization dedicated to sharing their resources in both computing and communications technology. Launched in 1965, the Interuniversity Communications Council, known as EDUCOM, aimed for collaboration in "all information processing activities . . . as examples, computerized programmed instruction, library automation, educational television and radio, and the use of computers in university administration and in clinical practice."[110] EDUCOM's first newsletter mirrored the paired interest in computing and communication with articles on time-sharing, computers on campus, and computer art.[111] By the end of 1965, twenty-five universities had joined EDUCOM with the goal of cooperation across their approximately one hundred separate campuses.

Precisely because EDUCOM initiated, collected, and reported on the latest in computing-communications resource sharing, it is a helpful lens through which to examine the rise of networking, a rise in which ARPANET was one network among many. During the decade after its establishment, EDUCOM served as a nexus of research and reporting on computing and communications technology, and its spring 1974 bulletin "Special Issue on Networking" captured the "remarkably swift progression of networking thought," including

Martin Greenberger's direction of its 1972–1973 seminars "to help identify the central issues in building and operating networks on a national basis."[112] Similarly, EDUCOM's invitation to Greenberger to manage those seminars represents another fruitful intersection because, as we have seen, Greenberger drove the discourse on computing utilities, computers and communication, and ultimately networking through his organization of conferences and publications during this time.

Although an EDUCOM working group called for a single national multimedia network in 1966, by 1974 the council reflected on the benefits of the nation's multiplying networks.[113] In July 1966, just a few months before Dartmouth dedicated its Kiewit Center, EDUCOM convened a summer working group on networks at the University of Colorado at Boulder. Its forty-two participants, representing universities and government, "proposed a multi-million dollar, multimedia pilot network that could eventually be expanded to interconnect all colleges and universities throughout the country."[114] Notably, those 1966 working group participants drew their inspiration from existing computing or communications networks, including those in California, Michigan, and North Carolina, at Dartmouth and MIT, and the Maine educational television (ETV) network. They noted that "the proliferation of regional or specialized systems . . . might enable EDUNET [their name for this national network] to become a 'network of networks,' as one expert suggested."[115] Eight years later, EDUCOM deemed that "networking actually proliferated across the country in a variety of operational—and thereby observable and testable—forms," which was preferable to "a single monolithic" network.[116]

By 1974, the computing networks included about thirty regional ones such as Dartmouth's (universities in Iowa, Texas, and Oregon anchored similar networks), Hawaii's ARPANET-linked digital satellite network, the Triangle Universities system in North Carolina, eight emergent statewide networks (including Minnesota's, which is the subject of Chapter 5), and the commercial networks offered by TYMSHARE, GE, and UCC, prompting the observation that ARPANET was "just one illustration of networking."[117] In its ALOHA System, the University of Hawaii overcame the restrictions of point-to-point wire communications by employing UHF radio broadcast channels and satellites to create a computer-communications network among the islands and with connections to ARPANET, the University of Alaska, NASA Ames Research Center in California, and Tohoku University in Japan.[118] North Carolina organized a network of forty-seven community colleges, colleges, universities, and technical institutes, as well as ten high schools, which received computing access via the Triangle Universities Computation Center of Duke University, North Carolina State University, and the University of North Carolina at Chapel Hill.[119] The creators of the Iowa Regional Computer Network, which served thirteen educational institutions from the University of Iowa, cogently observed, "From the start it has been clear that the Iowa Regional Network . . . was something more than what appears on paper—that is, an organization consisting of faculty and students, computer professionals, computer and communications equipment, and software. It has been evident that this was not a sterile utility, but rather a dynamic organism wherein each institution and its members help the other institutions."[120] I want to emphasize two points in their comment: first, they included faculty and students in the network. Their definition of a network included people (who were not computing professionals), as well as hardware and software. Second, they recognized the cooperative and collaborative nature of the emergent network, such that the sum of the network was greater than its parts. Recall that Kurtz had made a similar observation about the Kiewit Network, recognizing that the computing network created a commons, a community around computing.

Conclusion

Whether it was called a computing utility, an information utility, a single network, or a network of networks, the prospect of national networking propelled the development of academic and commercial computing-communications networks from 1960 through the early 1970s. During that time, many commentators also clamored for government involvement in such an endeavor, whether to nurture its development, regulate it, or put it to service for society. Most believed that computing could be a utility—one like electricity or water—something that the government regulated for the good of its citizens. Herein lay the promise of computing citizenship, participation in a networked world in which the nation, with its states and cities, encouraged access. In 1969, Kurtz declared, "Ways and means must be found for federal support for this new utility. There is a historical comparison between 'computer power' today and 'electrical power' some thirty years ago. Federal funds gave us light—now federal funds must give us knowledge."[121] Chapter 5 shows how one state, Minnesota, fulfilled this call. Starting with an early computing connection to Dartmouth College, followed by its own proliferation of time-sharing networks, Minnesota organized a thriving statewide time-sharing network by the mid-1970s.

5

How *The Oregon Trail* Began in Minnesota

THE VIDEO GAME *The Oregon Trail* was wildly popular during its heyday. It was inducted into the World Video Game Hall of Fame in 2016, hailed as the "longest-published, most successful educational game of all time."[1] One can still play it online, or purchase it for a smartphone or tablet. The kids who grew up playing *The Oregon Trail* during the 1980s and 1990s recognize it, cheer for it, and wax nostalgic about it.[2] The phrases "caulk the wagon" and "you have died of dysentery" (for that matter, the words "caulk" and "dysentery") penetrated popular consciousness and culture because of *The Oregon Trail*. Over the past decade, online outlets have regularly covered the history of *The Oregon Trail*, pieces (including a podcast) that play to the game's long-standing popularity and its high nostalgia factor.[3] These articles all tell a similar story, starting with the game's three student-teacher cocreators writing it in 1971 to enliven an eighth-grade history class that one of them was teaching. The rehearsed story then jumps to the game's resurrection for the Minnesota Educational Computing Consortium (MECC) in the mid-1970s, followed by MECC's partnership with a very young Apple Computers, and the dramatic spread of MECC's educational software on school-based personal computers during the 1980s.

Yet this conventional *Oregon Trail* wisdom misses some crucial details, overlooking the much longer and bigger history of people computing in Minnesota during the 1960s and 1970s. This chapter situates the history of *The Oregon Trail* and MECC into Minnesota's rich history of social computing during those decades. In 1967, eighteen Minnesota school districts formed a unique organization known as TIES (Total Information for Educational Systems) as a cooperative venture to provide educational and administrative computing to their students and teachers.[4] The TIES districts had been inspired by a 1965–1966 computing experiment at University High School (UHigh) in Minneapolis. The success of TIES propelled the creation of MECC in 1973.[5] During 1974–1975, MECC's statewide time-sharing system served 84 percent of Minnesota's public school students.[6] By 1978, students played OREGON, their beloved game *The Oregon Trail*, on the MECC network over nine thousand times per month.[7]

This chapter contends that TIES and MECC users engaged in social and creative computing practices that now feature prominently in contemporary American digital culture, including networked gaming, social networking sites, and user-generated content. Analyzing the growth of TIES and MECC illuminates the social and technical practices of networked computing that were distinctive to the networks' origins within education during the 1960s. The individuals who established TIES and MECC focused on schoolchildren and their teachers as users and innovators. The networks' employees, teachers, and students emphasized access, cooperation, and emotional and intellectual engagement.

They cultivated a community around their time-sharing networks, built around a social vision of collaboration and user orientation. For TIES and MECC users, computing became both participatory and personal. Highlighting the educational origins of TIES and MECC underscores the diversity of networks and social practices that have created our contemporary pervasive computing culture, but that we now associate exclusively with today's Internet. We must attend to the history of computing and networking from the ground up—from the user's perspective—to fully grasp the evolution of our contemporary computing culture.[8]

The focus on Minnesota is neither provincial nor coincidental. From the 1950s through the 1980s, Minnesota enjoyed a thriving economy based on computers.[9] At the time, no other region in the

United States could match Minneapolis–St. Paul for the concentration, size, and success of its computer industry. This regional business grew from the pioneering Engineering Research Associates, founded in 1946 and an early manufacturer of state-of-the-art computers. The Minnesota economy was anchored by the presence of major corporations Control Data, Honeywell, Sperry-Rand Univac, and IBM Rochester, but hundreds of other companies large and small supplied necessary parts, services, and know-how for the large computer companies. The Minnesota computer industry profoundly affected the culture of the state, especially in the Twin Cities area. Residents had a "computer identity," viewing themselves at the forefront of technology.[10] During the 1960s and 1970s, Minnesotans excelled at computing, from the boardroom to the classroom. This leadership extended beyond TIES and MECC. During the 1960s and 1970s, Minnesota-based Control Data substantially contributed to the development of the interactive PLATO (Programmed Logic for Automatic Teaching Operations) system at the University of Illinois at Urbana-Champaign, another time-sharing system on which thousands of users created individualized, interactive computing.[11]

TIES and MECC built their networks around time-sharing; the technological form of time-sharing and the social organization of these networks went hand in hand. For TIES and MECC users, personal computing and networked computing were inseparable. The students and educators of TIES and MECC embraced time-shared computing for entertainment and personal information processing. They created programs to compose music and to process their income taxes; they savored simulations such as MANAG, the business game, and SUMER, in which they ruled an ancient civilization. They shared their programs, as well as their burgeoning computing expertise, through the TIES and MECC networks. TIES and MECC users made computing their own—they made it personal—in many ways: they had one-on-one interaction with the teletypes; they computed for productivity, for communication, and for fun; they experienced emotional engagement and sociability with their computing, and they cultivated computing communities.

This chapter documents the movement for participatory computing in Minnesota by first tracing its origins in a computing experiment conducted at UHigh in Minneapolis starting in 1965. Several UHigh teachers then employed social movement organizing tactics to call for more computing in the classroom. The resulting network, TIES, employed an intensive communication strategy to encourage participatory computing. While TIES enriched its network, other educational systems—including the Minneapolis schools, in which *The Oregon Trail* premiered—arranged their own time-sharing access or time-sharing networks. The TIES network's success with user-generated content and a software library, together with the proliferation of other computing systems, propelled the statewide network, MECC. MECC replicated the communications strategies of TIES and, more importantly, developed a statewide telecommunications network to support widespread interactive, individualized computing.

UHigh: The Experiment

Dale LaFrenz has characterized himself as a math teacher rather than a mathematician, but he is, at heart, a salesman extraordinaire.[12] Throughout his career, he has sold the idea of computing in the classroom to peers, administrators, and students in Minnesota and across the United States, starting at one high school. LaFrenz was one of four new teachers in the mathematics department at UHigh in Minneapolis for the 1963–1964 school year.[13] The College of Education at the University of Minnesota had established UHigh in 1908 as a place to conduct research on teaching and learning, to train teachers, and to experiment with novel curriculum approaches emerging from the college.[14] UHigh was, in short, a laboratory school, and both its teachers and students were guinea pigs. UHigh prided itself on being at the forefront of innovation, and during the 1963–1964 school year, the new math teachers searched for a novel educational experiment for themselves and their students.

Inspired by Minnesota's thriving computer economy, LaFrenz and his colleagues aimed to "bring the computer into the classroom."[15] They were curious about whether the computer could be used effectively in an educational setting, and they soon learned of a promising way to study this: the Dartmouth Time-Sharing System (DTSS).

Kemeny agreed that the UHigh teachers and their students could join the DTSS, provided that UHigh cover the extremely high cost of long-distance telephone service from Minnesota to New Hampshire.[16]

The Dartmouth system greatly appealed to the math teachers at UHigh because they could install a teletype at the school, thereby providing their students with hands-on access. LaFrenz and his colleagues considered student use of the teletype (not the mainframe computer itself) a form of computing. LaFrenz emphasized, "We put the teletypewriter in the classroom. That's really where the whole computer in the classroom started."[17] Moreover, the teachers presented teletype usage to their students as computer usage.[18]

For 1965–1966, the first year of the experiment, UHigh sought and received $5,000 in funding from the GE Foundation because GE had provided the computer for DTSS. Most of that grant was applied to long-distance service to Hanover. The experiment included seventh-grade classes taught by Larry Hatfield, ninth-grade classes taught by LaFrenz, and eleventh-grade classes taught by Thomas Kieren.[19] LaFrenz's students employed the computer in learning about the order of mathematical operations and the evaluation of numerical expressions.[20] Hatfield's seventh graders in the experimental group used the computer to learn about exponential numerals, while Kieren's eleventh graders studied linear and quadratic functions with the computer.[21] The teachers were united in their conviction that the "computer could serve to provide problem-solving experiences for *all* students in grades 7–12," not just those who were mathematically talented or those in eleventh and twelfth grades.[22]

Various forms of publicity drew attention to—and support for—the UHigh endeavor. In a *Minneapolis Tribune* article titled "University High Is 'Lab' for Curriculums That Are 'Far Out,'" the reporter Richard Kleeman cited the teletype as his first example of the school's "far-outness," its experimental and innovative approach to education, and the article featured a photo of math teacher Pamela Woyke and two students using the teletype.[23] UHigh's mid-1960s promotional book prominently featured the computing experiment.[24] This thirty-two-page booklet advertised the school to prospective students and their parents, but it also commemorated the

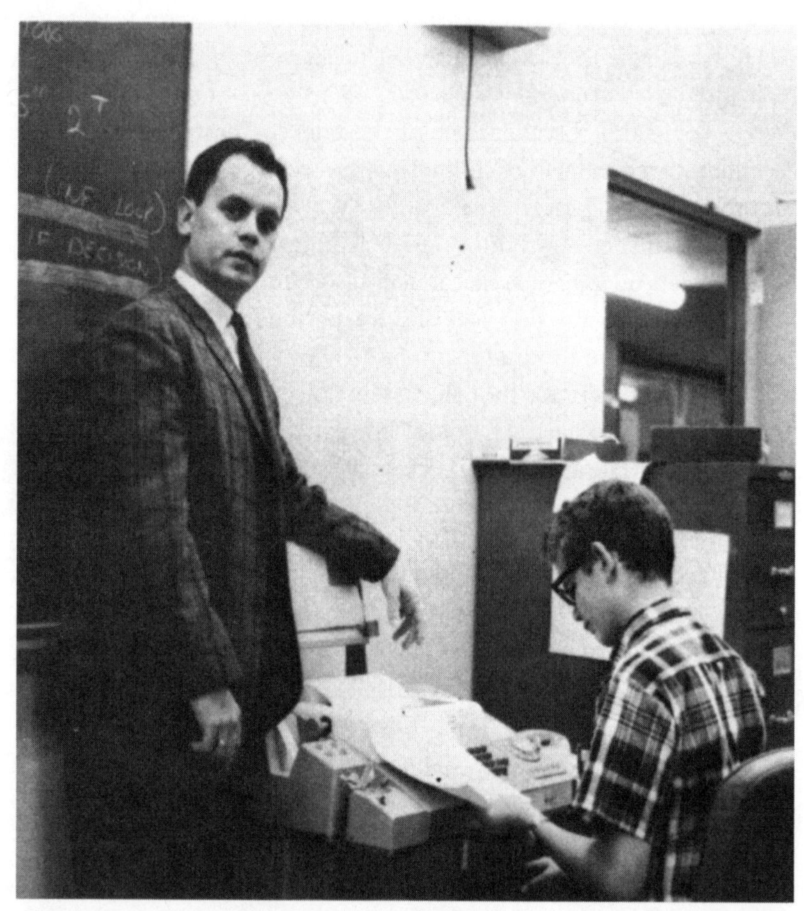

Dale LaFrenz works with a student at a teletype terminal at UHigh. UHigh students and educators initially gained time-sharing access by connecting to the DTSS (also known as the Kiewit Network). Thus began a Minnesota computing experiment that culminated in a statewide time-sharing network a decade later. Courtesy of University of Minnesota Archives, University of Minnesota–Twin Cities.

school for current students and served to introduce the school to other educators, politicians, and policymakers in Minnesota and beyond. In the introductory pages, the booklet highlighted the computer experiment as one of three examples of the newly prioritized goal of curriculum research, and the section on mathematics featured a large photo of a student operating the teletype, with additional information about the project.[25]

The student editors of the 1966 UHigh yearbook splashed a bold headline across the two pages about the math department, "Use of Computer Enriches Teaching of Modern Math," and the accompanying article described the experiment.[26] The 1967 yearbook also trumpeted, "Math Computer Project in Second Year," and featured a quarter-page photo of LaFrenz and a student with the teletype.[27] The math section of the 1968 yearbook headlined, "Computer Aids Students in Understanding Theories," and the article described how students often used the computer during their lunch hour, which evinced the growing popularity of computing.[28] LaFrenz attested to competition for the computer, explaining that "the whole school was trying to share it."[29] Students featured the computer experiment in their newspaper as well as their yearbook. Among articles about the chorus caroling downtown and the ski club planning a trip, the December 1965 issue of the UHigh *Campus Breeze* reported, "Computer hook-up opens door for math teaching experiment" and detailed the Dartmouth connection, time-sharing, and the nature of the experiment.[30] Less than a year later, the experiment made front-page student news; the reporter described how the computer experiment necessitated a revision of many of the school's math courses to accommodate the computing experience.[31]

The *Campus Breeze* covered computing uses beyond the math department, documenting the growing popularity of the time-sharing teletype. In November 1967, the "A Glance Around" section of the paper revealed plans for a "Computer Dance," elaborating, "Students would fill out computer cards telling about themselves, and the computer would select couples."[32] Computing merited front-page status again in April 1968; the article "Computer Game Intrigues Students and Teachers" detailed a management decision-making game played on the computer.[33] Teams of juniors and a team of math teachers competed against each other to outsell their competitors in this reproduction of "the market situation of real beer companies."[34] The head of the social studies department explained, "The game simulates reality. . . . [It] allows you to experiment with something you just couldn't do for fun in real life."[35] In fact, the computer had been used beyond the math curriculum almost from the beginning; UHigh teachers computed to teach physics and to provide demonstrations to outside math teachers and other groups.[36] As computing became integrated into UHigh's physics courses and its summer math-science courses, the experiment leader Johnson proclaimed, "The student enthusiasm in these [classes] was exceedingly high."[37]

From UHigh to TIES: The Network Grows

LaFrenz and his fellow UHigh teachers eagerly spread the word about computing in the classroom, especially after time-sharing costs dropped dramatically. In 1965, the Minneapolis-based Pillsbury Company became the first commercial venture in the nation to purchase a GE-635 computer, and Pillsbury soon opted to install and sell time-sharing on its GE machine.[38] UHigh switched its teletype connection from Dartmouth to the Pillsbury subsidiary Renown Properties in February 1966.[39] Members of the UHigh group recognized that this local time-sharing option eliminated the long-distance costs associated with their computing model, and they aspired to expand the student computing experience. The partnership between UHigh and Pillsbury underscored another dimension about time-sharing networks, namely, that their reliance on telephone service made for local or regional networks. LaFrenz reminisced,

> The five of us began evangelizing the use of the computer in the classroom and what we were doing and time-sharing. We began going to the Minnesota Council of Teachers of Mathematics [MCTM] meetings to "sell" our idea. Pretty soon there was quite a cadre of people in the Twin City area who had convinced schools to buy teletypes and hook up and start using the computer in the classroom.[40]

Indeed, the MCTM provided the UHigh group with a ready-made and receptive network through which it could propagate its idea. During 1965–1966, David Johnson served as coeditor for the MCTM newsletter, providing the crucial connection between the UHigh experimental group and a large statewide network of educators.[41] As editor, he arranged for two MCTM articles publicizing the UHigh computing situation. Johnson clearly intended to convince others to join the computing crusade. In his own piece, he concluded, "In view of the tremendous impact of computers on our society it is with great excitement and expectation that the department is conducting this research."[42] Johnson tapped Larry Hatfield to publish in the May 1966 issue of the MCTM newsletter, and Hatfield also advertised the UHigh experiment.[43] Johnson soon promoted computers in the classroom to a national audience at the 1967 conference of the National Council of Teachers of Mathematics.[44]

LaFrenz, Johnson, and Hatfield's efforts to convince other schools about the importance of computing in the classroom paralleled the tactics of their 1960s social movement contemporaries. In her now-classic essay, "On the Origins of Social Movements," Jo Freeman argued that a social movement required (1) a preexisting communications network that was (2) readily co-optable, as well as (3) additional organizing work to disseminate the new idea.[45] Applying that analysis to Minnesota computing highlights the methods by which LaFrenz and his colleagues laid the groundwork for a TIES social network. They promoted their idea through the preexisting communications network of the MCTM. They deployed both the newsletter and meetings to recruit other educators (teachers and administrators) to the cause of instructional computing. The additional organizing work would come through the development and growth of TIES. The individuals associated with TIES exhibited the characteristics of a social movement: they were conscious of a shared enthusiasm for computing, they demonstrated a missionary impulse to spread their message, and they mobilized many others to pursue a common cause, culminating in MECC.[46]

During the spring and summer of 1966, while the UHigh teachers wrapped up the second year of their computing experiment and traveled around the state spreading the gospel of computing, Twin

Cities educators learned of another University of Minnesota School of Education computing endeavor. The School's Educational Research and Development Council (ERDC), which provided research and consulting services for metropolitan school systems, distributed numerous questionnaires in over twenty school systems to survey teachers and administrators about potential uses of computers in their schools.[47] The responses to those questionnaires shaped the effort that would become known as TIES. TIES began as a cooperative venture among eighteen school districts in the Minneapolis–St. Paul metropolitan area, and the system aimed to provide both administrative and instructional computing services for over 130,000 students.[48]

The first step in creating TIES was the recognition by the ERDC that Minnesota laws encouraged a particular form of local government uniquely matched to the strengths of 1960s computing. Minnesota's Joint Exercise of Powers Act enabled political subdivisions of the state, such as cities, counties, or school systems, to come together as a single entity. That single entity could then exercise any of the powers afforded to its member units, such as purchasing or hiring.[49] Thus, ERDC executive secretary Van Mueller recognized that the Twin Cities metropolitan school systems could form a cooperative venture enabling them to pay for—and share—the resources of a large mainframe computer. This was an expense and undertaking that almost no single school district could afford on its own. Mueller and the ERDC were aware that other school districts around the country, and even in Minnesota, used some data processing services for tasks like scheduling or payroll. However, the ERDC envisioned a system serving member schools that went beyond data processing to encompass people, training, application development, and student computing.[50]

The results of the ERDC questionnaire indicated strong support for a cooperative data processing project, and by January 13, 1967, eighteen school districts had adopted resolutions to join the Minnesota School Districts Data Processing Joint Board, which became known as TIES.[51] This joint board sought federal government support for its project under the Elementary and Secondary Education Act of 1965 Title III.[52] The application for Title III funding evidenced three critical features of this particularly Minnesotan project. First, teachers were considered partners in the development of the technological system and consulted from the outset.[53] Second, the area was permeated by its own high technology culture surrounding the numerous local computing companies.[54] Finally, the university also acted as a key contributor, as it had for UHigh.[55] Indeed TIES manifested the same constellation of forces as UHigh: teacher expertise and innovation, bolstered by the university, and situated within a high-tech hub.

After securing their initial Title III funding, the school districts of TIES aspired to self-sustainability.[56] The system planned to operate on per-pupil membership fees from participating districts, so growth decreased costs for all involved. As more school districts joined TIES, the number of students served increased, and the fixed costs associated with owning and operating the time-sharing system decreased by being spread among more districts. Here, the employees and supporters of TIES adopted the organizing techniques of the movements of the 1960s to spread their message around the Twin Cities and across Minnesota. They used three key organizing techniques to accomplish their goals: meetings, local coordinators, and newsletters. They worked to convince existing school districts of the value and utility of their investment in computing, and they strived to persuade other school districts to join in their computing collaboration. To maintain existing members and recruit new ones, TIES staff organized their activities to instill a sense of accomplishment and pride for TIES efforts, and to meet the computing demands of their large constituency.

The TIES technological network was simultaneously a social network, and the social movement was also first and foremost a social network. The TIES staff organized numerous school visits, meetings, and training sessions to inform and energize their constituents about the potential of their information system. These face-to-face encounters constituted a TIES effort at mobilization. The meetings commenced shortly after TIES began operations; each of the twenty-one member school districts received a visit from joint board personnel between Thanksgiving and Christmas 1967.[57] These meetings among TIES staff and member teachers, administrators, and students continued on a frequent basis over the next five years; this frequency highlighted their value to the TIES organization.[58] In fact, the layers and diversity of groups organized is striking. The June 1968 issue of the *TIES & TALES* newsletter detailed a meeting of the joint board, with members drawn from each school district, as well as workshops for the Educational Information System (EIS) coordinators from each district, a Technical Committee meeting, and a computer concepts seminar attended by representatives from all member districts.[59]

A key aspect of TIES success was this attention to individuals, this in-person component. Part of the TIES mission was to familiarize teachers and administrators with computing and everything it could do for them. Executive Director Thomas Campbell, Assistant Director Jerome Foecke, and the others recognized that the people in their system were just as important as the machines, and they encouraged frequent contact to ensure that questions were answered, concerns were allayed, insights were shared, and milestones were celebrated.

Closely related to this communications strategy of frequent meetings was TIES's requirement that each member district designate an EIS coordinator, who liaised between the district and TIES. One of the coordinators, Irv Bergsagel, reported that he essentially served as a "communications link."[60] Bergsagel realized that he and the other EIS coordinators played a vital role in the TIES network. They kept information and ideas moving within their time-sharing computer network and their computing community. The EIS coordinators embodied all of the local places within the TIES community, as well as the spaces in between and imagined on the computer network.

TIES launched its *TIES & TALES* newsletter in September 1967 to apprise members of its activities.[61] TIES staff published the newsletter several times during the academic year. They distributed a total of eighteen *TIES & TALES* newsletters during their first five years of operation.[62] The initial TIES staff carefully attended to geography in the TIES newsletter articles, thereby creating a network across member districts and beyond for their readers. Aware that their project spanned multiple counties and, at that point, intangible planning for a futuristic-sounding computing system, the staff sought to firmly ground their activities for their constituency. They conveyed this message on the second page of their first newsletter, stating that when deciding on their "long range base of operations . . . a number of items such as short and long range communication networks, highways, and geographic distribution of member districts . . . will be taken into consideration."[63] The attention to geography extended beyond Minnesota. As reported in the newsletter, TIES staff communicated with colleagues in New York, New England, California, Michigan, and Oregon.[64] They welcomed visitors from Michigan, Palo Alto, and the University of Southern California.[65] They reported on an educational technology conference in Scotland and on student computer simulations in Westchester County, New York.[66] TIES staff used their newsletters to create a network of individuals, schools, and school systems across Minnesota linked by time-sharing, but they also forged connections and a sense of belonging in a network across the United States and beyond. The TIES staff filled each newsletter with the promise of computing.

The software banking and other network effects that TIES facilitated underscore the importance of understanding TIES as people-focused and community-based—that is, as a social network. The use of time-sharing in TIES member schools exploded during 1970, paralleled by the emergence of TIES as a software repository. During the 1970–1971 school year, over twenty-six thousand students used the TIES teletypes.[67] The newsletter explained, "As more and more teachers and students become involved with the BASIC [programming] language and the use of the computer, additional programs are generated and additional uses of the devices are developed."[68] The special structure of TIES enabled this phenomenal growth of usage and programs: if one student in one member district wrote a program, he could save it to TIES's computer library, where it could be called up, used, and modified by another student or teacher in another TIES member district.

TIES students and educators interacted with their terminals in myriad ways. Students eagerly played games such as CIVIL (a Civil War emulator) and MANAG, even outside of class.[69] Older students created entertaining and informative demonstrations about computing for younger students. Another group of students recorded a video about how information was processed from the teletype to the computer, and Linda Borry, a teacher, programmed the computer to compose music.[70]

This rapid growth prompted TIES to create a new role in each member district, that of the terminal supervisor. The terminal supervisor supported the use of the computer in the classroom and encouraged use of the software library.[71] With the terminal supervisor, TIES effectively instituted a help desk role in each member school system in 1970. The development of this role evidenced TIES's status as a software creator and its awareness of the importance of supporting users. The TIES staff recognized that maintaining and energizing their existing user base was just as critical as recruiting new member districts.

In fact, the terminal supervisor role was one piece in the complex system that TIES administrators and educators developed to regulate use and access around the increasingly popular time-sharing system. One middle school teacher "established a procedure for students to receive a computer operator's license similar to a driver's license," and other schools followed suit.[72] By October 1972, there was enough competition for teletype time that winning a personal user ID with unlimited access time for one month was a valuable prize.[73]

A final emblem of the expanding enthusiasm for instructional computing was the launch of the *Timely TIES Topics* newsletter in September 1972.[74] The newsletter was devoted to sharing student computing news and programming ideas, and it demonstrated that computing in the classroom was becoming institutionalized. The teletypes were integrated into the classroom spaces of TIES member schools, and teletype usage was embedded as an option for thousands of students. Because *Timely TIES Topics* regularly included contributions from TIES teachers and students, and because those contributions always included school location information, the newsletters together with the time-sharing system represented a distributed yet connected network. Readers belonged to a community that was connected by telephone lines and computers, and by the possibilities and passions of computing.

The TIES leadership used both *TIES & TALES* and *Timely TIES Topics* to present their vision of a better future by showcasing TIES as a role model for other computing ventures. Indeed, this projection of TIES as a thought and action leader worthy of international attention unified the staff's attention to geography, community, and training. For example, when commenting on his attendance at the National Council of Teachers of Mathematics regional meeting in Cedar Rapids, Iowa, LaFrenz—now a math consultant for TIES—reported that "current computer problem solving efforts by TIES member schools are leading the way in computer assisted mathematics."[75] By 1970, the ambition and scope of TIES earned attention from other states and even other countries. Various national publications including *The Nation's Schools*, *Educational Technology*, and *Computer Management* featured articles about TIES, and the *British Financial Journal*, comparable to the *Wall Street Journal*, dispatched reporters to Minnesota to investigate the project.[76]

Campbell and his TIES staff touted this coverage, but they also highlighted Minnesotans' appreciation of just how singular their project was. Several TIES school district coordinators attended the national Association for Educational Data Systems conference in Miami, which provided an opportunity to compare their system with those of Wisconsin, Oregon, and other districts and states. The EIS coordinator, Bergsagel, reported that TIES computing was far more sophisticated than comparable systems and was "a leader in the field."[77] Another coordinator, Jerry Seeman, emphasized the status of TIES as a telecommunications network and the cooperation of those involved with the network.[78]

Indeed, that telecommunications network set the TIES program apart. Although the network was established partially for centralized and cost-effective handling of administrative functions such as payroll and scheduling (information processing), the network also facilitated interaction among students, educators, and schools as they increasingly shared their "computers in the classroom" experiences. As the TIES headquarters staff actively crafted this image of leadership, they emphasized the system's unique characteristics: "movement into the field of instructional applications of the computer, and the regional-cooperative approach," which included extensive training and on-site coordinators.[79] Just three years after the project officially launched, TIES celebrated itself as "a model educational information system" and "a 'front runner,'" attracting notice from "scattered points around the nation, plus England and Sweden."[80] Campbell and his colleagues proudly proclaimed that TIES was "a model for legislative action" for "a statewide network of regional information systems."[81] In fact, at the urging of the governor, legislative action would soon result in the formation of the MECC, another joint-powers organization that fostered rich time-sharing experiences.

Interlude: All Paths Lead to Minnesota, and then along *The Oregon Trail*

Threads from Chapters 1–4 became interwoven in Minnesota. Bob Albrecht, the BASIC (Beginners' All-purpose Symbolic Instruction Code) evangelist, worked for Control Data Corporation in Minnesota in the early 1960s. He preached the gospel of time-sharing to the young math teachers at UHigh, including Dale LaFrenz. They called Kemeny and Kurtz and arranged a connection to the DTSS in its early days. The long-distance telephone connection between Minneapolis and Hanover was extraordinarily expensive; however, a local alternative soon appeared.

The Pillsbury Company (familiar to many for its Doughboy mascot), headquartered in Minneapolis, was an early adopter of GE's time-sharing computers—the GE computers that had been built on the expertise of Dartmouth students. UHigh bought Pillsbury computing time, as did other schools in Minneapolis and its suburbs. In January 1967, the *Minneapolis Star* reported that six Twin Cities suburban high schools purchased Pillsbury-GE computing time. "Students at the schools . . . responded by developing computer programs their teachers say are amazing. . . . [One] program permits students to play baccarat [a game] with the computer."[82] A few months later, three Pillsbury representatives traveled to Hanover to participate in Dartmouth's Secondary School Project training on computing in the classroom, led by John Warren from Exeter, one of the schools in the Secondary School Project network.[83] When Digital Equipment

Corporation (DEC), Hewlett-Packard (HP), and their competitors introduced time-sharing minicomputers, Minnesota schools turned to that lower-cost computing option. The Minneapolis Public Schools arranged for their own HP time-sharing computer, with a terminal in every school, as did the TIES network.[84] Jean Danver, who had coordinated the Dartmouth Secondary School Project, forged another connection when she visited TIES as an HP employee. She presented a workshop to TIES staff, "Computer Assisted Instruction," as part of her computer-curriculum creation responsibilities at HP.[85]

It was in this high-technology state, already humming with people computing, that students explored *The Oregon Trail* in 1971. Don Rawitsch, Bill Heinemann, and Paul Dillenberger had been student teachers in the Minneapolis Public Schools when they envisioned using the school's time-sharing terminals to enliven their American history courses.[86] They brainstormed a game in which students would manage their resources, hunt, caulk their wagons, and battle weather and disease as they journeyed westward from the Missouri River to the Pacific Ocean. In December 1971, the now-famous game *The Oregon Trail* premiered in a Minneapolis high school. Students typed "BANG" to hunt, and they answered questions about how much to eat, whether to stop for rest at a fort, and how to cross a river. And they were utterly enthralled. But student teaching ended in December, and *The Oregon Trail* went dormant until MECC provided a new network on which students could travel the trail.

A Statewide Network: MECC and the Challenges of Multiple Stakeholders

By the early 1970s, TIES was not the only organization offering interactive computing experiences to students and educators in Minnesota; it had achieved its goal of becoming a model for others. Several similar projects had been successfully installed, and this proliferation of computing attracted the attention of Minnesota governor Wendell Anderson. A cooperative of private colleges, public community colleges, and public state universities formed the Minnesota Educational Regional Interactive Time-Sharing System (MERITSS) in 1971. This extensive time-sharing network originated from a computer housed and managed at the University of Minnesota in Minneapolis.[87] Similarly, Mankato State College hosted a time-sharing network for southern Minnesota known as the Southern Minnesota School Computer Project.[88] Thousands of people were computing. These were not just students doing preprogrammed drill exercises or even programming for their math assignments. Students were figuring out how to score volleyball games and swim meets, and they were playing simulation games such as CIVIL, in which the player chose Civil War battle strategies and soldier conditions.[89] High school athletic coaches were scheduling tournaments, calculating player statistics, and even determining their scouting choices via teletype terminals on time-sharing systems.[90]

These ventures called attention to the costs and inequalities of educational computing in Minnesota, and in July 1972, the Governor's Joint Committee on Computers in Education convened to review the state's computing activities, ultimately resulting in the establishment of MECC.[91] This committee aimed to satisfy several goals. John Haugo, who had previously worked at TIES and who acted as a consultant to the state on the development of MECC, observed that the rapid growth of educational computer use across Minnesota necessitated an effort for statewide planning and coordination.[92] The committee desired this coordination because they wanted to provide students who lived outside the Twin Cities metropolitan area with access to computing; that is, the committee sought to equalize educational computing opportunities for "outstate" students and educators.[93] The government also aimed to cap ballooning educational computing expenditures.[94] The demonstrated success of existing cooperatives such as TIES and MERITSS inspired the Governor's Committee to hope that a statewide venture could also thrive.

Plans for MECC proposed to unite the computing needs of the K–12 schools, the community colleges, and the state universities under one organization, and, as a result, members of those different communities questioned MECC from the outset. Moreover, MECC was essentially a product of the state, and the government involvement also invited criticism. Whereas TIES, MERITSS, and the Southern Minnesota School Computer Project had developed locally, from the ground up, MECC originated as a top-down government mandate.

The responses to the draft report on MECC, collected during the autumn of 1972, reflected the multifarious concerns. Members of the university community believed that the computing needs of higher education were fundamentally different from those of K–12 education and that combining them would be "unwise."[95] University officials understandably wanted to protect the interests of their own faculty, administrators, and students, including the seasoned users who had employed the university's computers or MERITSS for work or recreation. They feared that MECC would be dominated by either the interests of K–12 education or those of the state government, and they suggested substantial amendments to the draft report as a condition for their support of the statewide computing consortium.[96]

Representatives from TIES and the statewide K–12 public schools also expressed concerns about MECC. The May 1973 issue of *TIES & TALES* featured an editorial by Executive Director Campbell, the first editorial to appear in the five-year history of the TIES newsletter. Campbell feared a loss of local school district control. He exhorted his members "that local school districts need to express to their legislators and to the Commissioner of Education . . . that the needs, requirements and direction of management systems for instruction, administration and research activities is not wrested away from control of the local school districts."[97] This concern about state control paralleled that expressed by university representatives. And just as members of the university community worried that their needs would be subordinated to those of K–12 education, Campbell and his TIES colleagues worried that their K–12 needs would be subordinated to those of higher education. The official response on behalf of the elementary, secondary, and vocational schools echoed the TIES response.[98] The K–12 educators and administrators felt that MECC could deprive them of their powers of decision making and of providing input.

The Governor's Committee addressed the issues raised by MECC's proposed constituencies by revising the Joint Powers Agreement under which MECC would operate, most notably by adding an addendum with significant language around users. The addendum, the last four pages of the agreement, presented the "MECC Basic Principles of Organization and Operation."[99] The Governor's Committee included these "fundamental" principles to address concerns about state control and about the balance of decision-making power; many of the principles centered on the "user," which they defined as "the systems and institutions of education which use services of the proposed consortium."[100] These principles proclaimed a bill of rights for the users, mandating that "the governance of the consortium will be under the control of the users," and "the needs for services will be defined by the users."[101] The principle that was to be repeatedly invoked in the early years of MECC declared that users could reject MECC services if the services were found somehow lacking: "No educational user of computer services and/or facilities shall receive, as a result of joining the proposed consortium, less service or less adequate service than needed and previously available through institutional and system resources."[102] This strong language of users' rights quelled the concerns of both the university and the K–12 school districts, including the TIES districts. The Joint Powers Agreement was signed by the four Minnesota educational agencies, and MECC officially commenced operations on July 1, 1973.[103]

Minnesota's educational institutions endorsed MECC because its constitution embraced users' rights. This attention to needs and preferences of Minnesotans as technology users attested to the penetration of interactive computing in the educational culture; however, this user orientation also posed challenges for MECC in its early years. Although the Joint Powers Agreement defined the "user" as "the systems and institutions of education," that is, the school districts or the university, most people involved with MECC understood "user" in a different way. They considered the many individuals—the faculty and students—who were doing some form of computing as the users with rights.[104]

During the autumn of 1974, LaFrenz, now the MECC assistant director for instructional services (previously at UHigh and TIES), endeavored to formalize this understanding of users' rights with a MECC User's Association. He organized meetings with representatives from school districts, TIES, and MERITSS, and he solicited feedback on drafts of the proposed constitution from educators around the state.[105] The Articles of Organization, adopted in 1975, declared, "There are two categories of membership: Faculty/Staff and Students. Any person who is a bona fide *user* of the MECC Instructional Time Sharing System qualifies for membership."[106] Thus, students and educators across Minnesota who used time-sharing under MECC's auspices believed that the consortium was accountable to them. But these users were far from a homogeneous group. The concerns that emerged during the MECC proposal process revealed that MERITSS and university time-sharing users sought to protect their experiences of interactive computing, just as TIES users sought to protect theirs. That is, MECC emerged into a technological environment in which there were existing groups of users with varying experiences and expectations of interactive computing, and there were potential users with still different goals and needs. MECC had to service all of these different users, who clearly had different ways of experiencing and valuing the MECC system. A teacher who had spent five years with TIES was, quite simply, a very different user than a student in rural Minnesota who experienced time-sharing for the first time under the system that MECC would build.

During its inaugural 1973–1974 year, MECC drew heavily from Minnesota's existing abundance of computing resources, including human resources, and the MECC staff implemented the techniques of meetings, newsletters, and coordinators that had contributed to TIES's success. Indeed, the early years of MECC attested to the entrenchment of interactive computing in Minnesota. The first three MECC assistant directors—Dale LaFrenz, Dan Klassen, and John Haugo—had been associated with TIES. LaFrenz had worked as a TIES instructional consultant from 1968 through 1970, Haugo had been the coordinator of educational research for TIES, and Klassen taught TIES summer workshops.[107] LaFrenz and his colleagues then organized numerous training sessions for teachers throughout Minnesota. These workshops, dedicated to introducing teachers to computers in the classroom, operated with the "goal of providing an opportunity for every school district to send at least one person to a workshop."[108] By July 1974, MECC had introduced nearly four hundred teachers to the possibilities of interactive computing.[109] In addition to using the TIES technique of meetings as part of its technological system, MECC also deployed local and regional coordinators and newsletters.[110]

MECC's primary focus during the 1973–1974 school year was the extension of time-sharing services to the outstate regions of Minnesota. It relied on existing, successful time-sharing systems, including TIES. It truly built on the computing knowledge developed in Minnesota over the past decade. The time-sharing systems it used encompassed TIES, MERITSS, the Minneapolis Public Schools, Mankato State College, Bemidji State College, and St. John's University, and it endeavored "to minimize changing a user from his current system."[111] Over the course of the 1974–1975 school year, MECC's statewide time-sharing system utilized five HP 2000 minicomputers, one Univac 1106 mainframe computer, and one Control Data Corporation 5400 mainframe computer.[112] Together, these minicomputers (not to be confused with microcomputers, as early personal computers were known) provided about 450 ports, or telecommunications entry points to the computers, accessed by approximately 800 terminals across the state.[113] However, making this plan work, involving numerous different systems, was not simply a matter of installing teletype terminals in school districts located in remote regions of the state.

The MECC staff's commitment to a software library was notable. They recognized that they could build on the work already accomplished in TIES, MERITSS, and other Minnesota time-sharing systems by converting existing software programs to a single MECC statewide system. For the 1974–1975 school year, during which the MECC statewide system included several different computers, the MECC staff labored "to provide as much of a common library as possible on each of the systems."[114] Their efforts for providing a common library included practice conversions from one computer system to another.[115] MECC aspired to harness the potential of this intellectual property by making it easily accessible to users around the state.

MECC's greatest achievement in creating a statewide instructional time-sharing system entailed the development of a supporting statewide telecommunications network. Individuals using teletypes to interact with time-sharing computers moved their data over telephone lines, and the cost of telephone time for calls beyond a limited local area was quite expensive. For example, Hibbing, Minnesota—one of MECC's school districts located about two hundred miles north of Minneapolis—was also about one hundred miles east of the nearest time-sharing computer at Bemidji State College. The MECC staff worked with the telephone companies of Minnesota to develop cost-effective means of connecting districts like Hibbing with remote time-sharing computers. One component of MECC's solution was the use of multiplexors, which were "communications devices that concentrate[d] many calls across one line to the computer."[116] In October 1974, when the network was 90 percent complete, MECC had established thirteen multiplexors around the state to reduce telecommunications charges.[117] Furthermore, the MECC staff worked with the telephone companies to install toll-free lines (at the time, telephone lines accessed by dialing an area code of "800") for "very remote schools."[118] Thus, the MECC staff, with the cooperation of the Minnesota telephone companies, adapted existing technologies for new purposes to implement their statewide time-sharing system. In this case, the statewide computer consortium acted for multiple districts across the state and therefore merited far more attention from telephone companies (and other businesses) than a single school district would.

In addition to exercising their purchasing power with the telephone companies, the MECC staff also worked with teletype businesses to extend instructional time-sharing beyond the Twin Cities metropolis. MECC negotiated a "cost beneficial arrangement" whereby MECC became the seller, or provider, of teletypes to school districts. Here, too, the consortium relied on bulk purchasing to benefit member school districts. Minnesota schools could purchase a popular teletype model at a discounted price.[119] The $1,270 price was attractive and attainable to numerous Minnesota school districts, and within four months, MECC had sold about 160 of these teletypes.[120] MECC also promoted a statewide teletype maintenance agreement through Minnesota-based Tele-Terminals, Inc. This contract allowed school districts to receive maintenance and service calls for their teletypes—regardless of whether they were purchased through MECC—at a discounted rate.[121] Thus, MECC encouraged school districts throughout the state to put computers in their classrooms by reducing the actual cost of obtaining and servicing the requisite teletype, and by minimizing the decision making associated with an individual school district purchasing its own teletype, finding a time-sharing provider, debating whether to enter a maintenance contract, and wondering how to actually use time-sharing.

MECC created a network of networks through innovative communications solutions and business negotiation, and by building on the extensive foundation of existing Minnesota time-sharing. This network of networks enabled thousands of students and educators across Minnesota to program and personalize the computers. Prior to the 1974–1975 school year, the Minneapolis–St. Paul metropolitan area accounted for the overwhelming majority of classroom computing in Minnesota school districts. Before MECC, only 14 percent of Minnesota students with access to instructional time-sharing were outside the Twin Cities metropolitan area. Once MECC implemented a statewide time-sharing system in 1974–1975, that number tripled to 46 percent.[122] During the first year of its statewide time-sharing system, MECC served 84 percent of the public school enrollment in Minnesota.

Just a few years later, during the 1977–1978 school year, students participated in computer simulations like OREGON in 42 percent of Minnesota public school courses.[123] By this time, *The Oregon Trail* had become quite popular throughout Minnesota. MECC had hired Rawitsch, one of the game's original developers, in 1974, and he resurrected the game, adding it to MECC's library. Students (and teachers) tapped into that library as they worked at problem solving and instructional games in 39 percent of their computer-related courses. And students enjoyed leisure time—time to pursue their own interests and programs—with the MECC time-sharing system in 32 percent of those courses.[124] An average day during the 1977–1978 year featured over five thousand user sessions.[125] MECC had achieved its goal of making interactive computing readily available for Minnesota's educators and students.

The Walker Art Center in Minneapolis celebrated the flourishing culture of computing in education in its exhibit *New Learning Spaces and Places*, on display from January 27 through March 10, 1974 ("Front Matter—Exhibit Catalog," *Design Quarterly* 90/91 [January 1, 1974]: 1). The learning station pictured here featured teletype terminals connected to a time-sharing network on which visitors could compose music or play a game of pool. Another station included teletypes on which visitors could play computer resource management simulations including CIVIL and POLUT ("Exhibition Floor Plan," *Design Quarterly* 90/91 [January 1, 1974]: 13–14; Ross Taylor, "Learning with Computers: A Test Case," *Design Quarterly* 90/91 [January 1, 1974]: 35–36). Another popular station included the interactive flat-panel plasma display touch screens of a computing system called PLATO (see Chapters 6 and 7). Photograph by Eric Sutherland for Walker Art Center.

Conclusion: The Bug in BAGELS

During the winter of 1974, interactive computing thrived in Minnesota schools. For example, students in TIES member schools played YAHTZE on their time-sharing terminals, a version of the classic dice game Yahtzee written at a TIES school by student teacher David Auguston.[126] Linda Borry, the teacher who programmed the TIES computer to play music, now worked on the TIES instructional staff, and she solicited help with an ongoing problem. Borry reported, "It has been brought to our attention that there is a bug in the BAGELS program which periodically causes it to print out incorrect clues. Can you help us find this bug?"[127] Borry knew that her computing community would help resolve the problem. Similarly, Tom Mercier, a wrestling coach at TIES member Lakeville Junior High School, impressed his colleagues with his computer prowess. He programmed the TIES time-sharing computer to calculate all of the pairings for Lakeville's fifth annual invitational junior high wrestling tournament, involving nine teams and 158 wrestlers. Mercier's program saved significant time during the meet and became part of the constantly growing TIES library as a resource for others.[128] Meanwhile, LaFrenz, who had started his journey at UHigh and worked with Linda Borry at TIES, diligently worked to build MECC's staff, create a telecommunications network crisscrossing the state, and share his zeal for computing with educators around the state.

Focusing on the network developed from UHigh through MECC reveals the spirit of collaboration that animated individuals like Borry and LaFrenz. Indeed, TIES cultivated people as the crucial component of a vibrant information network. TIES employees and affiliates also organized their venture as a social movement, using newsletters, meetings, and local coordinators to mobilize Minnesota communities and spread the gospel of computing. Moreover, LaFrenz, Borry, and their colleagues worked with—and pushed—the limits of 1960s and 1970s computing systems in an effort to connect those computers with many different people (not just tech-savvy individuals) as soon as possible. They did not dwell on the limitations of time-sharing; rather, they maximized computing opportunities. They built collaborative, user-focused, educational-driven computing networks around their time-sharing systems. Sometimes the Minnesotans improved the technology, but they always prioritized increasing access. And in the process, these leaders and their many users redefined computing.

Hundreds of thousands of Minnesota students and educators made computing their own. For these TIES and MECC users, computers no longer loomed as a specter of science fiction, nor were they the province of only scientists and engineers. Large corporations and research universities did not have a lock on regulating computer access. Instead, for the participatory Minnesotan computing community, computing became individualized and interactive. Computing became accessible and personally meaningful—as a way to accomplish homework, play games with friends, find a date, or calculate taxes owed.

Studying TIES and MECC demonstrates the importance of unconventional settings for the history of networks and the history of computing. It seems that education has been overlooked largely because contemporary Americans imagine that technological use in the classroom was narrowly circumscribed. This underestimates the creativity and agency of users in shaping technologies. Examining TIES and MECC also illustrates the value of looking beyond the technical implementation of a network, for TIES and MECC thrived based on their social practices as well as their time-sharing capabilities. Similarly, the history of TIES and MECC underscores the human labor required to produce networked computing. For networks, computing, and networked computing, we must move beyond details of devices and protocols to consider the history of human actions and activities in creating applications, ascribing value, and determining social practices. The students and educators of TIES and MECC cultivated participatory computing, and their legacy informs networked computing today.

6

PLATO Builds a Plasma Screen

In 1963, Maryann Bitzer created a new way for nursing students to learn how to treat heart attack patients. The students were in their first year at the Mercy Hospital School of Nursing in Urbana, Illinois. Bitzer's course took them a short distance to the Coordinated Science Laboratory on the flagship Urbana-Champaign campus of the University of Illinois. There, the students began their lesson by sitting down at a terminal with a television screen and a set of keys, all of which were connected to a mainframe computer. The computer provided the material for the lesson, including a film that each student watched on-screen and questions to answer afterward. Bitzer had introduced these nursing students to a time-sharing computing system known as PLATO (Programmed Logic for Automatic Teaching Operations).[1]

For Bitzer and her students, the practice of reading to acquire new information was familiar; reading on a screen was not. The screens to which these 1960s students were accustomed were those on which filmstrips were projected in their classrooms, and those of the televisions—pieces of furniture—in their homes. The PLATO screen was more the size of a textbook, a personal size. The practice of answering questions aloud or in writing to demonstrate understanding on a particular subject was familiar. The practice of pushing a button to answer a question, using some keys wired together, was not. Perhaps most unfamiliar was the computer in the role of teacher; Bitzer's PLATO nursing program provided information on request and tested students on their learning. The nurses were not just learning how to treat cardiac arrest; they were also learning how to interact with an individual computer screen and an individual computer terminal. They were crafting personal computing.

Maryann's husband, Donald Bitzer, had invented the PLATO system at the University of Illinois in 1960. PLATO began as a project to investigate how computers could be used in education, and the first version was a one-user "teaching device" with a screen and keyset, wired to a mainframe computer.[2] By 1969, the PLATO system featured thirty-five terminals, complete with screens and keyboards, connected to a dedicated mainframe Control Data Corporation 1604 computer. Students and faculty could use twenty of the thirty-five terminals at the same time because PLATO had become a multiple-user time-sharing system.[3] During the 1960s, under the auspices of the PLATO project, Bitzer also shepherded the development of flat-panel plasma display screens and touch screens. These screens, along with programmable keysets, were installed on the nearly one thousand terminals that supported a nationwide PLATO network of students and educators by 1975.[4]

Focusing on the hardware of the PLATO system, and the ways in which it was developed to meet users' needs, changes our narrative of personal computing innovation. The educational context had nurtured PLATO during its first decade, and it enabled the development of a large-scale network of personal terminals. The thousands of students and educators using PLATO did not have to pay for their access; they were computing citizens, not computer consumers. My focus on the relationship between PLATO users and their personal terminals recasts existing histories of personal computer hardware that privilege Silicon Valley. I am not interested in who first had the idea for a personal computer or who first created a particular piece of hardware. Rather, I aim to understand under what circumstances personal terminals were developed, and how and by whom they were used. This chapter and Chapter 7 answer those questions.

Student at PLATO I terminal with individual screen and keyset. Photo courtesy of the University of Illinois at Urbana-Champaign Archives, image 1140.tif.

The first section of this chapter analyzes the genesis and early years of the PLATO system. The next section, on the innovation of the plasma screen, contends that the expansive PLATO IV network of the 1970s was possible precisely because it was pitched for American public education. The third and final section examines a 1974 PLATO user's manual as the lens through which to reconsider a Silicon Valley legend.

The Birth of PLATO

Donald Lester Bitzer had the good fortune of attending college and graduate school at the University of Illinois during the 1950s. He had grown up in Collinsville, a small city in the southwest part of Illinois, from which one can see the Gateway Arch in St. Louis across the Mississippi River. During the first half of the decade, while Bitzer majored in electrical engineering, the Cold War escalated on the battlefields of the Korean Peninsula. The United States fought to prevent the spread of communism from China across Asia. The Cold War shaped Bitzer's path at the university too.[5]

While Bitzer pursued his doctorate in electrical engineering, he worked at the university's classified Control Systems Laboratory (CSL). The university had established the lab in 1951, during the Korean War, so faculty could pursue military-sponsored research in Urbana-Champaign. During the 1950s, the faculty and graduate students affiliated with the CSL pursued projects that were military priorities during the Cold War, including rockets, radar display screens, the propagation of radio signals, and an air-defense system. Bitzer's dissertation research reflected another military priority: antenna systems. Antennae are essential to broadcasting and communications. Some antennae translate the electrical signals of radio, television, telephone, or radar systems into radio waves so the signals can travel thousands of miles, and other antennae receive those radio waves and convert them back to electrical signals.[6]

Many of the projects at the CSL, including Bitzer's, employed the Illinois Automatic Computer, or ILLIAC. When the ILLIAC became operational in 1952, it joined only a handful of other electronic digital computers in the United States. Researchers programmed the ILLIAC to analyze antenna and radar patterns, to evaluate atomic blast effects, and to find the stress points in materials used to build bridges.[7] In his dissertation research, Bitzer programmed the ILLIAC to identify antenna beam patterns, to calculate integrals, and to check experimental flight data. He was, in short, very familiar with the resources of the Control Systems Laboratory and the ILLIAC.

PLATO's genesis, then, occurred at a Cold War research laboratory, at one of the few universities that had a computer at the time, designed by an electrical engineer with military funding. In 1959 the Control Systems Laboratory changed its name to the Coordinated Science Laboratory (keeping its CSL acronym) to reflect a shift from classified to unclassified work, and away from an overtly military focus. As part of this transition, the lab director and his associates sought a way to experiment with computers for education. Bitzer had defended his dissertation on December 10, 1959, and during the spring of 1960, he was searching for his next project. The CSL director's description of the challenge of creating computing for students fell on Bitzer's primed ears.[8]

Bitzer's experience at the CSL informed his expansive sense of the possibilities for PLATO, and he and his small team built a functional PLATO I system between June and November 1960. He knew that his fellow CSL researchers, in their work on aircraft radar tracking, had used some high-quality electrical output storage tubes manufactured by Raytheon. Those tubes could be used together with a television screen to create a graphical display for the student user.[9] Bitzer also envisioned how the powerful digital processing of the ILLIAC, with which he had firsthand experience, could be used to create "the best" experience for each user.[10] He jotted down these ideas in his lab notebook in June, then he began building.[11] Bitzer focused on the electrical engineering tasks, namely, wiring all the necessary hardware for the system. This included engineering the use of the Raytheon storage tubes for the television screen, wiring the screen itself, building a keyset created from individual push buttons, and integrating a slide projector for use with the screen. Peter Braunfeld, a mathematician, managed the ILLIAC programming tasks, such as programming the ILLIAC to communicate with all of the hardware components, and programming what was called the "logic" for the PLATO system. The "logic" addressed how instructional material would be displayed to the student user, how and when the computer would communicate with the student, and how and when the student could communicate with the computer. A handful of other CSL staff members assisted Bitzer and Braunfeld in bringing PLATO to life in six months.[12]

When Bitzer and Braunfeld introduced PLATO to the wider world by publishing an article and giving a conference presentation about it, they eased their audience into a sense of familiarity with their new technology by referencing the tools of the schoolroom—books and blackboards.[13] As the diagram on the following page illustrates, a student user sat in front of a display screen (the TV display) and a keyset.

The screen displayed text, images, or a combination of text and images, or it showed short films. The "electronic book" and the "electronic blackboard" provided material for the screen, for the con-

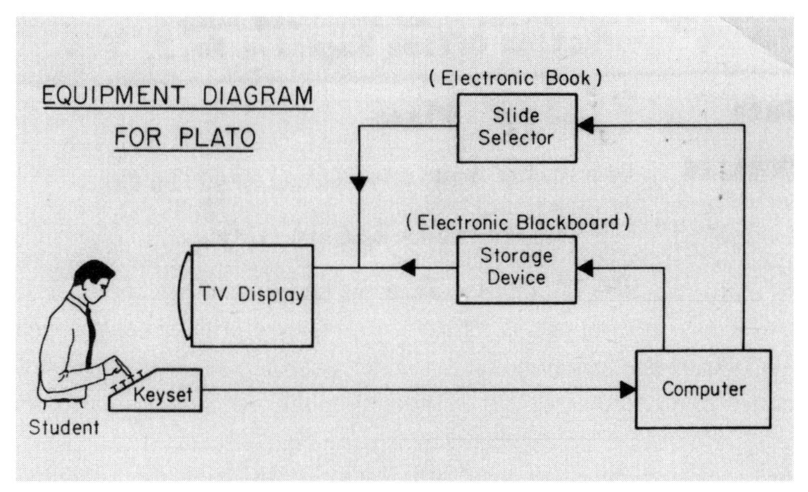

Diagram of PLATO I equipment, with "electronic book" and "electronic blackboard." Photo courtesy of the University of Illinois at Urbana-Champaign Archives, image 3304.tif.

sumption of the student user. Here, Bitzer and his colleagues referenced the traditional classroom essentials of books and blackboards to demonstrate continuity with 1960s classrooms, and to subtly underscore how computing could transform those classrooms.[14] Their "electronic book" was a set of prepared slides that displayed text and/or images, and the slides were projected onto the TV display. The "electronic blackboard" referred to the space on the screen where the computer or the student user could make characters (mainly letters and numbers) appear. The blackboard name conveyed the concept of writing and erasing: the computer was programmed to write a question on the blackboard (to make the characters of a question appear on the TV display), and the student used the keyset to write an answer on the blackboard (to make the characters of the answer appear on the TV display). After the student answered a question correctly, and before the computer showed new instructional material, the computer erased the blackboard, clearing both the question and the answer. The authors framed the interaction between the student and the computer as one of communication.[15] The student used the keyset to communicate with the computer, and the computer used a closed-circuit television to communicate with the student. Bitzer and his PLATO team emphasized the student user's activity and choices when the student was using PLATO, thereby highlighting the reciprocity of the system.

A PLATO student user interacted with the TV display and the keyset as the hardware of the PLATO system, and encountered the "programmed logic" as the software of the system.[16] The PLATO logic coordinated the interaction of the computer with a student via a "main" teaching sequence and a "help" sequence. Once the student was seated in front of the TV display and keyset, the student viewed slides from the lesson on-screen. After viewing and reading the material on the first screen, the student pressed the CONT. (continue) button on the keyset to continue to the next screen. At some preprogrammed point, the computer displayed a question about the material on-screen; that is, the computer asked the student to answer a question. The student then used the keyset to answer the question. If the student entered the correct answer, the computer displayed "OK," and the student proceeded with subsequent material in the lesson until the next question. If the student entered an incorrect answer, the computer displayed "NO," but it did not display the correct answer. The student could attempt to answer the question multiple times. At any point in the process of viewing instructional material on-screen or answering a question, the student could also press the HELP button on the keyset and enter a help sequence that provided additional information or examples to clarify the lesson material. Within the help sequence, the student might also encounter questions from the computer, for example, queries that broke a mathematics problem down into component parts. This particular programmed logic remained in place, with some additions and revisions, throughout the 1960s. Bitzer and his colleagues had initially characterized PLATO as a "teaching device" or a "teaching machine," thereby conveying that their project was, above all, an experiment in education. During these early years of the system, the team conceived of their users only as consumers of instructional materials, that is, as simply students.

Bitzer's drive to create a working PLATO system within months reflected his obligations to the military and to the university. The CSL still received its funding from the United States Department of the Army (Signal Corps and Ordnance Corps), the Department of the Navy (Office of Naval Research), and the Department of the Air Force (Office of Scientific Research, Air Research, and Development Command) through a Signal Corps contract.[17] Like at MIT and Stanford, scientific researchers at the University of Illinois benefited from substantial military funding during the Cold War. But rather than submit projects for approval, CSL employees could commence work on a project, then communicate the details of that project to the military via the lab's quarterly progress reports. The military then reviewed projects at the end of the year to allocate funding for the next year. Bitzer recognized that the sooner he could produce a prototype of value to the military, the more likely he would be to receive funding. Thus, Bitzer and Braunfeld introduced the project as a "teaching machine" in the March, April, and May 1960 CSL quarterly progress report.[18] They declared that it would have applications to military education for training in projects such as SAGE (Semi-Automatic Ground Environment), the extensive computer-based radar system that coordinated the American military response in the event of an attack from the air (missiles, planes) by the Soviet Union. Bitzer and Braunfeld demonstrated PLATO to the military in late 1960. Bitzer recalled, "They saw it and they wanted more of it. They decided the military may be the largest educational facility in the world. . . . They considered this important military research work and therefore would allow their money to go to support it."[19]

At the same time, Bitzer and the CSL had to inform the university of potential financial benefit for the university based on the development of PLATO. Bitzer and Braunfeld, listed as coinventors on the form, reported in their 1961 patent disclosure to the university that they considered PLATO's hardware and software for possible lucrative technology transfer from the university to a corporation. The project's focus on education distinguished it from many other military-sponsored university research projects during the Cold War era. A computing laboratory centered on students necessitated very different collaborators and questions than those labs (at MIT or Stanford) investigating air defense or nuclear weapons. In particular, Bitzer's eagerness to work with scholars in the humanities and education transformed the project from a "closed world" into an open one.[20]

That Bitzer was an engineer building an education system propelled the expansion of PLATO. The team required educational materials for their teaching machine, but they were scientific researchers, not teachers. Bitzer and his CSL colleagues needed educators and subject specialists to create PLATO lessons and be willing to use the system with their students. The research and development team publicized their project across campus, around the state, and to CSL visitors to recruit university faculty and students, as well as K–12 teachers and administrators, to the project. As early as 1961, the team reported a compelling public demonstration in the town of Allerton Park, thirty miles away from the Urbana-Champaign campus. The researchers set up a PLATO keyset together with an "ordinary television set" as the PLATO terminal in Allerton. A local television station broadcast PLATO images to the ordinary television set; that is, a local station broadcast the electronic book from the ILLIAC to Allerton. Local telephone lines connected the Allerton keyset to the computer on campus, and the telephone lines transmitted signals for the electronic blackboard through combinations of five tones. Those tone combinations enabled the computer to pose questions to the audience at Allerton, and for the demonstration student to communicate answers to the ILLIAC using the keyset. The audience was enthralled.[21] These public demonstrations, conducted on campus and beyond, generated support for PLATO in the wider university and Champaign-Urbana communities. They also helped Bitzer form strategic research partnerships, such as those with University High School, the University of Illinois Committee on School Mathematics, and Mercy Hospital School of Nursing.[22] The team also publicized the system—and marked their academic territory—by publishing numerous timely articles about it and by presenting at conferences, such as the 1961 Western Electric Show and Convention in San Francisco and the 1961 Conference on the Application of Digital Computers to Automated Instruction.[23]

Bitzer's research alliance with his wife Maryann, using PLATO to deliver instructional material to nursing students, was precisely the kind of partnership that transformed PLATO from a teaching machine to a republic of computing citizens. Maryann devised her experiment for her master's thesis in nursing, and she emphasized her use of PLATO as a "simulated laboratory."[24] Maryann cited two motives for her study: the dearth of trained nursing instructors and the tremendous educational value for nursing students of working with actual patients. PLATO helped on both fronts.

The first-year students at Mercy Hospital School of Nursing used PLATO to learn how to care for a patient with a heart attack. Maryann chose a striking topic. Patients would never want student nurses experimenting on them in the life-and-death situation of a heart attack, but PLATO enabled those students to experiment on a virtual patient. After a short introduction to the system and the terminal, the student nurse watched on the individual screen a three-minute film that portrayed a conversation between a doctor and a man "to convey an image of a real patient."[25] The film ended as the man experienced chest pains, and the student entered the simulated laboratory on-screen. Here, the student could experiment with administering various forms of nursing care or different drugs, such as nitroglycerine, and observe the effects of these actions on the virtual patient.[26] Maryann praised the "creative thinking" and "self-discovery" that PLATO fostered by enabling students to learn at their own pace, to ask questions, to experiment, and to observe results.[27]

While the student nurses learned about caring for heart attack patients, they also had to gain familiarity with the keyset. The keyset featured only nine usable buttons, one labeled "Lab" and the rest numbered one through eight. The "Lab" button functioned like a "home" key; pressing it always returned the user to the starting slide of the PLATO lesson. Students could not type questions or answers in familiar English syntax. Rather, they had to hunt for the numeric code that corresponded to a particular phrase, such as "check conditions of patient" or "give or change drugs," and then use the numbered buttons on the keyset to enter that code.[28] Moreover, the options for the keys numbered one through eight changed, depending on where the student was in the PLATO program. If the student pressed an "illegal button," the keyset beeped.[29] While the students worked with PLATO, Maryann watched their performance on television monitors. She observed that "minor system failures . . . tended to create anxiety in the student," but overall the students responded very favorably.[30]

Considering the evolution of the PLATO keysets demonstrates the team's early ideas about learning, and how they adapted keysets to work around the limits of their system hardware. The first PLATO keyset had only sixteen keys.[31] Furthermore, the keyset was not a keyboard in the contemporary sense of the word. Rather, it was simply sixteen push-button keys grouped together. Bitzer elaborated, "They didn't make keyboards in those days. You made your own. You used push-buttons and you arranged them."[32] Because the team needed the keyset to display ten digits, punctuation marks, and special commands, they created two sets of labels for the keyset: one label that appeared under normal light, and another label that appeared under ultraviolet light. Bitzer and his researchers then installed an ultraviolet light over the terminal and switched the terminal lighting between normal and ultraviolet to make visible the different keyset labels, depending on the student's place in the lesson.[33] The keysets featured buttons unique to PLATO: CONT., REV., JUDGE, HELP, AHA!, CALC., and a button for "refreshing" the screen.[34] After viewing a particular screen, the student pressed the CONT. (CONTINUE) button to go to the next screen. REV. (REVERSE) took the student back to the previous screen. After inputting an answer using the keyset, the student pressed JUDGE to check whether the answer was correct. Pressing the HELP key moved the student away from the main teaching sequence and into the help sequence.

At any point in the help sequence, if the student realized how to solve the problem, the student could press the AHA! key to RETURN to the main sequence. In other words, rather than completing the entire help sequence, the student could use the AHA! key as an escape back to the main sequence. Depressing the CALC. key communicated to the ILLIAC to display a calculator on-screen for use by the student. The RENEW key addressed the limits of the Raytheon television screen. Bitzer reminisced, "Storage tubes fade with time. You sit there and after about three minutes and you figure out what you want to do [with PLATO], you'd look up and there's this big white spot right in over everything you wanted to see."[35] Bitzer and his

An early PLATO keyset. Note the "AHA!" key in the column of keys on the right side of the keyset, located third from the top. Image courtesy of the Charles Babbage Institute, University of Minnesota Libraries, Minneapolis, Minnesota.

team recognized how frustrating that disappearing screen would be to their users, and they wanted to ensure those users "wouldn't be disappointed."[36] They implemented the RENEW key to refresh the storage tubes. "It would redo your whole screen for you real quickly," Bitzer noted.[37] From the start, Bitzer instilled in his team consideration for their student users, and attention to the relationship among their system's hardware, software, and users.

After Bitzer had demonstrated that the ILLIAC could be used to create a teaching machine, and that the students could learn from that teaching machine, he expanded the system via time-sharing. By the summer of 1961, his team had connected two PLATO terminals to the mainframe computer and ensured that two students could use those two terminals at the same time.[38] Although the two-terminal time-sharing system (known as PLATO II) was barely operational in the summer of 1961, the team started planning a multiple-terminal system.[39] The fall 1962 progress report communicated plans for a total of thirty-two all-new terminals connected via time-sharing (this multiple-terminal time-sharing system was known as PLATO III).[40] All of this occurred very early in the days of time-sharing. Fernando Corbató and his colleagues at MIT first discussed their Compatible Time-Sharing System at the Spring Joint Computer Conference in 1962.[41] John Kemeny and Thomas Kurtz led the way for Dartmouth Time-Sharing in 1964. But in 1961 in Illinois, from the perspective of each user sitting at a PLATO terminal, the terminal—and the computer powering that terminal—was the student's alone.

The Plasma Panel: A Solution Looking for a Problem

As PLATO expanded, Bitzer played to his electrical engineering strengths by focusing on the system's hardware, and he encouraged research into the development of a plasma storage tube as part of the project. During the winter of 1962–1963, the experimental use of PLATO III (the multiple-terminal time-sharing PLATO) was just beginning. Bitzer and his team wanted to build a total of thirty-two PLATO III student terminals, but for each terminal, they had to manually wire the circuits connecting the Raytheon storage tubes, the cathode-ray display tubes, the keysets, and the computer.[42] Ultimately, the team required three years to build and program twenty student terminals.[43] Bitzer recognized that the high cost of the Raytheon storage tubes would limit the number of terminals that could be built, so he sought "a less expensive replacement for the present storage tube system" by attempting to create a plasma storage tube.[44] Plasma is an ionized gas consisting of a roughly equal number of (positive) ions and free (negative) electrons. The first experiments that winter investigated an electrical grid structure for stimulating the plasma. Thus, while he managed the growth of the PLATO III system for education, Bitzer also shepherded the research and development of a plasma physics project.

Over the course of 1963, the PLATO team began viewing the plasma project as not just a storage tube but also as a display tube, a crucial change of orientation. By the fall of 1963, they reported that a "plasma discharge display tube" would be "relatively inexpensive to manufacture" and, "being inherently a digital device, would display the computer information correctly."[45] They contrasted the digital plasma display with their current cathode-ray tube (CRT) display. The CRTs required digital to analog converters to take graphical input from the digital computer and display the output on the analog CRT. Because the images displayed on the CRTs faded over a few minutes, the CRTs also required sweep generators to maintain and refresh their graphical displays. The research group acknowledged that a "plasma display tube consisting of a large number of cells filled with an inert gas" had been proposed by Lier Sigler in the January 1963 issue of *Electronics*, but they hastened to describe their own achievement: they had designed a switching array for selecting and controlling a display device of 256 by 256 cells.[46]

The PLATO team described their plasma display as an "inherently digital device" because of how it was designed. The plasma display was essentially three very thin pieces of glass sandwiched together, with the plasma form of a gas such as neon or argon in between the glass pieces. The center piece of glass had a rectangular array of holes (such as 256 by 256 holes) cut out of the center. One of the outer glass panels featured electrodes running in horizontal lines, while the other outer glass panel had electrodes running in vertical lines, "such that each hole or cell [was] crossed by a pair of electrodes."[47] Voltage applied to a particular point in the array, or matrix, caused the plasma to produce a discharge of electricity at that point, which appeared as light. Thus, each cell in the matrix was either on or off, and could be represented as on = 1 and off = 0, making the plasma display a digital device.

During 1964, the plasma display researchers, especially Robert H. Willson, worked on the problem of controlling how cells in the array fired, or lit up. When they began their experiments with the plasma display, they arranged their electrodes within the tube, as part of the gas environment. However, they found that when they fired a large group of cells, adjacent cells (that should have remained off) also became illuminated. They puzzled over the "firing of adjacencies within the array" for months.[48] Finally, over the summer, Willson proclaimed "excellent success" with his solution of placing the electrodes on the outsides of the glass panels.[49] He also associated significant savings with his project, asserting, "This new system should cost less than one-twentieth of the present one."[50] His statement surely captured attention. Willson's only elaboration on this cost savings was that, "As each student site must have a memory device, the savings for even a small number of student sites can be large."[51] Having overcome the obstacle of the "firing of adjacencies" and having projected a huge cost savings, Bitzer, Willson, and the rest of the PLATO team endeavored to solve the one remaining problem that prevented them from building a "practical plasma display tube," that of charge buildup.[52]

The problem of charge buildup occupied Willson and the rest of the plasma researchers for the first half of 1965. They had to figure out how to modulate the voltages applied to the gas to differentiate between firing a cell from "off" to "on" and sustaining a cell in the "on" position after it had been fired. They also worked on the problem of lowering the firing voltage after a cell had been fired. They investigated tubes of different widths and hole diameters, as well as various gas additives, to differentiate firing and sustaining voltages.[53] With the help of two new team members who joined the PLATO plasma group later in the year, they solved the problem of the voltages.[54]

During 1966 and 1967, Bitzer and Hiram Slottow focused on building a fourteen-square-inch plasma display panel for a prototype next-generation PLATO terminal.[55] This meant a screen of about 3.74 inches by 3.74 inches, with a 512 by 512 point array creating 262,144 individually addressable points on the display.

Although scientists and engineers at other laboratories had been working on research related to plasma display panels, Bitzer and his colleagues knew their work was state of the art, and they staked their intellectual property claim through their conference presentations, publications, and, most importantly, a patent application. The Illinois researchers declared that they had "demonstrated" the plasma display panel's "soundness," while other laboratories merely "confirmed" the Illinois work.[56] For example, Bitzer corresponded with William Mayer at the Control Data Corporation research laboratories about Mayer's construction of a 132 by 132 plasma array.[57] In 1968, Bitzer and Slottow filed a patent application for a "Plasma Dis-

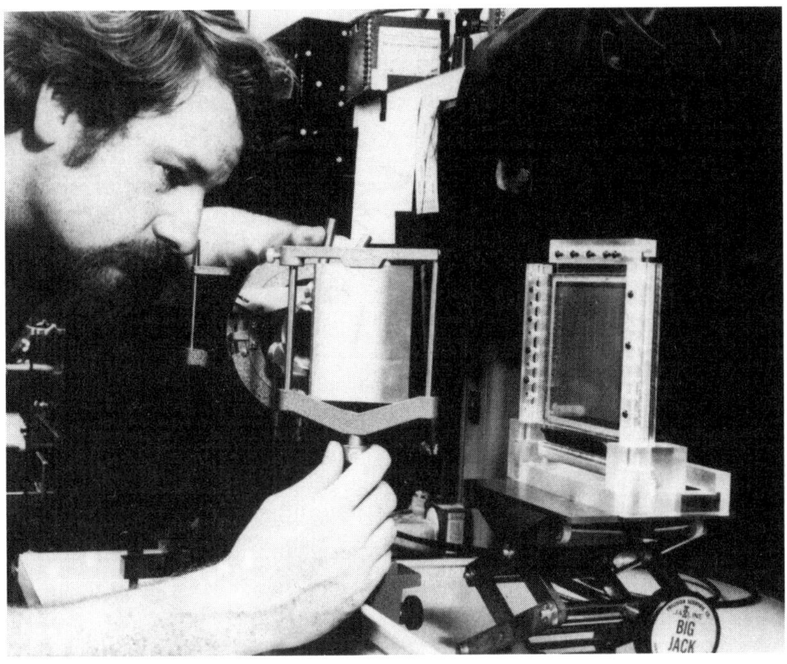

PLATO plasma display panel, circa 1966. Photo courtesy of the University of Illinois at Urbana-Champaign Archives, image 6443.tif.

play Panel Apparatus Having Variable-Intensity Display," with the patent to be assigned to the University of Illinois, and the patent was awarded in 1971.[58] The university profited from the patent by licensing production rights first to Owens-Illinois Glass Company and later to IBM.

Bitzer and his colleagues touted their plasma panel findings at conferences, including the 1966 Fall Joint Computer Conference in San Francisco, by enumerating the problems with CRTs. CRTs were expensive in and of themselves. They required digital-analog converters. They lacked an internal memory, and so the images on CRTs required constant regeneration. CRTs required high bandwidth, and therefore wide-band transmission lines were needed for images to be broadcast from a central computer over most distances. CRTs demanded high voltage to power them. CRTs with their associated memory tubes occupied a large amount of space relative to an individual workspace.

The PLATO group had developed a catalog of CRT problems such that for each problem, their plasma panel offered a tidy solution. Plasma displays responded directly to digital signals from the computer, so they did not require digital-analog converters. Plasma panels had their own memories, so they retained their images. They required neither external memory tubes nor constant image regeneration. The PLATO team estimated that thousands of PLATO terminals in a single community could each be connected by a single, inexpensive telephone line to a central distribution point. As for size, plasma panels allowed for variable screen sizes, and the panels themselves were quite thin (just three panels of glass).[59]

The PLATO team emphasized the range of options for both screen size and screen resolution with the plasma panel, and they showcased how various screens could be used beyond the educational realm. With CRTs, the density of points varied roughly with the size of the tube, but with the plasma panel, it was easy to add or subtract cells, to increase the density of cells, or to vary cell size. In 1966, they mentioned that a bank teller might have a display of 512 by 64, but "for the military command room, much larger displays are indicated."[60] By 1968, their projected applications included flat-panel televisions, aircraft instrument panels, and large wall displays for military war rooms or corporate boardrooms.[61] They pictured plasma display panels in banks, airline reservation centers, and corporate and university administrative offices. Bitzer and Slottow even predicted a world in which "thousands of people in classrooms and even homes will communicate simultaneously" using their system.[62]

The successful plasma-panel prototype turned Bitzer's attention to the economics of PLATO; he envisioned a dramatic expansion of the system based on the low cost of a plasma screen terminal. Histories of the plasma panel portray the device as a solution to an educational problem: it met "the need for a full graphics display for the PLATO system," or it was invented "to make it more comfortable for students working in front of computers for long periods of time, as plasma screens do not flicker."[63] In fact, the plasma panel became a solution waiting for a significant problem. It had been developed as a less expensive alternative to the CRT. But in 1966, the PLATO III system was running quite successfully with CRT displays, and

PLATO III would continue to grow over the next five years, with CRT displays. For PLATO III, the less expensive alternative to the CRTs was not necessary.

It is hard to imagine that PLATO IV would have been built without the plasma screen. Bitzer would have had a much harder time expanding the PLATO project with the economics of CRT displays. The cost of a PLATO III terminal with a CRT display in 1968 was about $5,000, and Bitzer projected that a PLATO IV system with thousands of terminals would decrease the per terminal cost to about $1,300.[64] The plasma screen enabled PLATO IV, and users revolutionized communications and computing with PLATO IV. It is difficult to overstate the importance of the plasma display for PLATO and computing.

PLATO matured against the backdrop of the 1960s as a "Golden Age of Education," when federal funding for education and scientific research ballooned.[65] The Soviet Union's launch of *Sputnik*, the first space satellite, in 1957 precipitated a wave of reforms in American education. The National Defense Education Act of 1958 legitimated sweeping federal aid to education by linking it to Cold War defense efforts, and it stipulated funding for technological and media services in schools.[66] Shortly thereafter, the National Science Foundation (NSF) supported the efforts of scientists and mathematicians to create new public school curricula. The Physical Science Study Committee and the Biological Sciences Curriculum Study brought together, respectively, elite physicists and biologists to overhaul American science education.[67] Other groups tackled chemistry, mathematics, and even social studies, producing the infamous "new math" curricula, as well as the "Man: A Course of Study" curriculum.[68] By 1977, nearly two-thirds of public school systems were utilizing at least one of these programs.[69] During the 1960s, scientists and engineers were widely esteemed, and they sought to remake American education according to their vision of science.

Events during 1967 prompted the NSF to directly support the use of computers in education. In February of that year, a prominent commission reporting to the President's Science Advisory Committee recommended extensive government investment in computing for education. The Panel on Computers in Higher Education, known as the Pierce Panel because it was chaired by John Pierce of Bell Telephone Laboratories, urged that all colleges and universities provide student access to computing, subsidized by the government. The Pierce Panel also proposed that the NSF and the Office of Education collaborate to encourage widespread student computing in secondary education.[70] President Lyndon Johnson heeded the advice of his Science Advisory Committee. In his February 28, 1967, speech to Congress, titled "Education and Health in America," Johnson announced his request for increased funding for the NSF, along with his directive to the NSF to partner with the Office of Education to investigate the potential of computing in education.[71] A few months later, the NSF established the Office of Computing Activities to oversee any federally funded research involving computing, including computing in education.[72]

Bitzer and the PLATO team began pitching the PLATO expansion in 1966, which culminated with the establishment of the Computer-based Education Research Laboratory (CERL) at the University of Illinois in January 1967 and substantial NSF funding the following year. At the 1966 Fall Joint Computer Conference, Bitzer referred to the "anticipated needs of the PLATO computer-based education system" and then mentioned hundreds of stations.[73] He was planning for something much bigger than his current PLATO III system. Bitzer described the team's shift to "emphasis on the development of a *large scale* computer-based educational system."[74] The university administration recognized "the importance of this large-scale concept to the future of the University of Illinois, and more broadly for the region and nation," and it created CERL to house the PLATO project.[75] As of January 1967, administrative and financial responsibility for PLATO shifted away from the science lab (CSL) and to the newly formed education research lab (CERL), and most of the CSL individuals who had worked on PLATO moved to CERL.[76]

During 1967, while PLATO III hummed along and users experimented with the new TUTOR programming language, Bitzer and other members of the PLATO research group sought financial support for the expansive PLATO IV system.[77] They contended that the United States spent $50 billion annually on education, but the nation still faced "vast unmet needs" in terms of both the quality and the quantity of education.[78] They lamented the challenges of serving a booming K–12 population while providing students with individualized instruction. The latter requirement especially resonated with the increased calls for equality in educational access resulting from the rights and protest movements of the 1960s, including the civil rights movement and the women's movement. As a solution, Bitzer proposed a time-sharing system with over four thousand terminals connected by standard telephone lines to one powerful mainframe computer. Each student terminal would, of course, feature a state-of-the-art plasma display screen to enhance the learning process. In March 1968, the NSF awarded CERL a $5 million grant (nearly $36 million in 2018 dollars) for the development of PLATO IV, a substantial amount supplemented by the state of Illinois.[79] The PLATO project continued to receive financial support from Mercy Hospital School of Nursing and Control Data Corporation. The Owens-Illinois Company, a major glass manufacturer, also provided funding, and it manufactured the plasma panels for the PLATO IV terminals.[80]

Building the new PLATO IV plasma-panel system while maintaining the popular PLATO III system occupied the team for five years. During the 1972–1973 school year, the team gradually placed the plasma-screen system into the hands of users. From September through December, the CERL researchers and Illinois educators worked closely and intensively to transfer PLATO III lessons to the new system, and to develop new PLATO IV lessons. During February, the team installed a new Control Data Cyber 73 computer and juggled all the associated hardware and software adjustments for their complex system. By that time, the team also supported 250 plasma terminals in nearly forty locations: twenty-five locations on the Urbana-Champaign campus and fifteen locations elsewhere.[81]

During 1973 and 1974, the number of PLATO terminals and users increased, and the project devoted significant resources to user-friendliness. Although the NSF had supported the endeavor since 1968, the academic years of 1973–1974 and 1974–1975 marked the formal demonstration and evaluation of the plasma-panel system for NSF.[82] Those two academic years witnessed dramatic growth in the number of lessons and amount of screen time. Prior to 1973, PLATO terminal usage by students had peaked at about twenty thousand terminal hours per year for 1969. That number declined from 1970 to 1972 while the team focused on building the plasma-screen system, and then terminal usage topped twenty thousand hours per year in 1973. PLATO IV exceeded expectations in its popularity. In 1974, students used the terminals for more than eighty thousand hours—a quadrupling of the previous record. As of July 1974, there were seven hundred PLATO IV terminals, a nearly tenfold increase over the PLATO III peak of seventy-one terminals. Certainly the dramatic increase in terminal hours per year can be attributed to the significant increase in number of terminals, but the plasma-screen terminals were also in use the great majority of the time. During 1974 and 1975, total terminal usage—by both students and "authors" (those creating programs in the TUTOR language)—ranged from a low of 40,000 terminal hours *per month* in December 1974 (during the semester break and holidays) to a high of 120,000 terminal hours *per month* for October 1975.[83] Managing this influx of new users and authors, and easing them onto the system, became a significant task for the CERL team.

Stealing Fire from the Gods . . . When the PLATO People Already Had It

During the summer of 1974, David Meller described the PLATO system in a handbook intended for new users. He highlighted the features of the remarkable PLATO IV terminal, and he employed humor to emphasize the novelty of these features. Meller cautioned, "Of the five senses, only tasting the terminal is not advised. Although smelling the terminal is permissible, it does not appear to be of any great benefit educationally."[84] Having captured his reader's attention, Meller continued, "You can, however, with rather pleasant results, see, hear, and touch it. The terminal will show you pictures, drawings, and writing; it will give you messages to which you can listen. It also responds to touching: it's a sensitive thing, you know."[85] For the students and educators at the University of Illinois who had used the PLATO system during the 1960s, being able to see the screen was not particularly noteworthy. They were accustomed to seeing graphics and watching movies on their CRT displays. Yet Bitzer and his

Using a PLATO plasma display terminal, circa 1972–1974. Photo courtesy of the University of Illinois at Urbana-Champaign Archives, image 1141.tif.

PLATO team had completely transformed the terminal from PLATO III to PLATO IV. That transformation encompassed the development of the built-in audio device and the touch-sensitive screen to which Meller referred in his introduction, a new programmable keyboard, and, of course, the plasma display that produced crisp and vivid graphics.

Although the plasma display was indubitably the heart of the new PLATO system, the team had eagerly announced additional new features in 1972. In March, Bruce Sherwood, who worked on the system software, updated the Association for Computing Machinery Special Interest Group for Computer Uses in Education.[86] He first outlined plans for several hundred plasma-panel terminals by the fall of 1973. Sherwood then described PLATO's random-access image selector, which could project any one of 256 full-color images onto the terminal screen in under 0.3 seconds, and he detailed a new random-access audio device capable of both recording and playing back sound. Sherwood introduced the touch screen as simply yet another useful computer accessory and explained that it consisted of a concealed row of sensors around the plasma panel, which formed a 16 by 16 grid. The computer then recognized when any one of those 256 regions was touched by a finger or other opaque object (like a pen), and it responded accordingly. The team anticipated that the touch screen would be especially useful for young children who could recognize images before words.[87] In his update, Sherwood also celebrated the system's flexibility. The keyboard could be completely reprogrammed for each new piece of software, if the software author so desired.[88] This meant that the key labeled "b" could be programmed to display (after it was pressed) an image of the atomic structure of boron, while "c" could be programmed to display carbon. Or the "b" key could be programmed to show a picture of a bear, while "c" could be programmed to show a picture of a cat. The programming language TUTOR, which was easy to use, facilitated this flexibility, as did the graphical display capabilities of the plasma panel.

Meller's user's manual provided snapshots of the rich, varied, interactive, and multisensory experiences that the citizens of PLATO's republic could enjoy. The manual was so popular that after its initial July 1974 publication, it was reprinted in December 1974 and then again in October 1975.[89] Meller began by proclaiming that for the user, the terminal was the user's personal computer. He declared, "The terminal means access to the PLATO system. For all practical purposes it is, to you the user, the whole PLATO system. . . . Whatever you're doing on PLATO, you are doing it at a terminal."[90] Meller then introduced users to the parts of the terminal, including the microfiche access, the plasma-panel screen, the power switch, and an old-fashioned escape key, the "error reset switch," for a "noncorrectable communication error."[91] He explained that each one of the 262,144 separate points on the plasma screen was a "dot of orange light," and that 17 dots of light formed the pattern for the character "c."[92]

Meller also reviewed the various user accessories, such as the random-access slide selector, the random-access audio device, and the touch panel. He observed that the touch panel was especially useful for young children or, for example, "asking a student to indicate where he would place a probe in the patient's brain."[93] In that case, knowing the verbal answer to the question demonstrated different (and perhaps insufficient) knowledge than knowing exactly where in the brain a region was located and pointing to it. Meller concluded his review of the terminal hardware by explaining the flexibility of the keyset. He explained that the keyset was a "device capable of far more than simply entering characters or a limited number of directives."[94] Rather, in various programs, depressing the "a" key could display the character "a" on the screen, or it could jump the user to a different lesson, or it could cause an alcohol group to be added to the chemical chain on-screen.

Amy Fahey was eleven years old when she encountered PLATO's memorable plasma panel and touch screen, playing "Speedway" on the system. Around 1974, she and peers from her Champaign school district regularly visited a PLATO lab, and they eagerly anticipated their computing time. Speedway combined car racing and arithmetic practice. At the start of the game, Amy chose her race course (Daytona, Indianapolis, Grand Prix, or Sebring), and the screen displayed a corresponding race track replete with grandstands, spectators, flags, and two race cars. Amy's race car sped ahead based on how quickly she answered basic math questions (like 6×6 or $12 \div 6$), and she competed against either the computer or her own previous best time (represented by the other race car).[95] She remembered "the excitement of looking forward to working on the computer and playing. . . . We even had touch screens installed later that year, which as an 11-year-old I'm sure I thought were the coolest."[96]

Meller's feast-for-the-senses handbook and Fahey's gaming enthusiasm undermine the most familiar story about where personal computers came from. That story usually features Douglas Engelbart, Xerox PARC, and Steve Jobs—Silicon Valley characters all. The very short version of that story is that Engelbart's 1968 demonstration of personal computing shaped the Alto personal computer built by PARC in the 1970s, and Steve Jobs stole the fire of the Alto—namely, the menus and windows on its screen and its mouse—and gave it to the people in the form of the Apple Macintosh. And then we had personal computers.

Engelbart is credited with inventing the mouse. During the 1960s, he worked at the Stanford Research Institute (SRI) to execute his vision of an individualized, interactive computer. He received funding from the Advanced Research Projects Agency (ARPA) to establish his own lab, the Augmentation Research Center. There, he and his employees developed a small computing network, known as the oN-Line System, NLS. Engelbart and his colleagues demonstrated the features of their system at the December 1968 Fall Joint Computer Conference in San Francisco, to an audience of about a thousand people. Over the course of ninety minutes, Engelbart used a terminal with a keyboard and a computer mouse, connected to a cathode-ray screen. A twenty-foot screen on stage showed the audience what Engelbart saw on his own screen. He created a document, used the mouse to move a cursor around the screen, copied lines of text, added graphics, and even brought in audio and visual elements. Moreover, the audience watched as researchers at Engelbart's lab, thirty miles away, also edited the text on Engelbart's screen via the networking capabilities of the NLS. Engelbart and the NLS received a standing ovation for their compelling presentation of personal computing.[97] But one member of the audience, Alan Kay, had already seen the prototype flat-panel plasma display that Bitzer had developed for the PLATO version of personal computing.[98]

Kay usually figures into the Silicon Valley legend because he saw Engelbart's demonstration, and the legend connects Engelbart's demonstration in a straight line to Kay's vision for personal computing when he went to work for Xerox PARC. Xerox, the photocopying giant, had established its Palo Alto Research Center (PARC) to conduct basic research. According to the legend, researchers at PARC created a magical personal computer, the Alto. The Alto was designed for personal use, and it featured a mouse-driven graphical user interface. Much has been written about how Xerox failed to market the Alto, as well as many of the other innovations that PARC researchers created during the 1970s.[99] The legend especially focuses on how PARC welcomed Steve Jobs to its campus in 1979, and how Jobs took the idea of a mouse, windows, and icons for Apple's Lisa and Macintosh computers. The technology journalist Steve Levy cemented the trajectory from Engelbart to PARC to Apple with his 1994 book *Insanely Great: The Life and Times of Macintosh, the Computer*

That Changed Everything. Levy characterized Engelbart's 1968 performance as "the mother of all demos," and wrote that "the next leap toward Macintosh would originate only a few miles from Engelbart's lab . . . known to computer-heads everywhere as PARC. It would become famous, but not quite in the way its parent company intended."[100] Apple struggled during the mid-1990s, only to stage a dramatic comeback with the return of Steve Jobs in 1997, the release of the iMac in 1998, and the introduction of the iPod in 2001. Levy's book enjoyed newfound relevance, and a legend was born.

Some have pointed out the problems with the legend, notably Malcolm Gladwell in the *New Yorker*.[101] Those who investigate the story observe that there are problems with the details, such as the fact that Apple was already developing a graphical user interface, even before Jobs's visit, or that it was actually two visits, not one.[102] The historian Alex Soonjung-Kim Pang, who produced and wrote the online exhibit *Making the Macintosh*, differentiated between invention and commercialization, and this is the theme that Gladwell picks up in his essay "Creation Myth." Gladwell argues that the legend is wrong because Jobs and Apple didn't want to just make a personal computer; they wanted to sell it. Or, as Gladwell puts it, "PARC was building a personal computer. Apple wanted to build a *popular* computer."[103] And this is certainly true. Jobs wanted to sell a lot of personal computers.

Gladwell is right insofar as pointing out that Jobs wanted to make popular computers, but adding the history of PLATO terminals to the mix points to a larger, more significant narrative. Jobs aimed to create a product that people would buy. He wanted to make consumers of personal computers. But at the University of Illinois, the people who used PLATO during the 1960s and 1970s accessed computing as members of a community, as part of a network. Students, educators, and community members sat in front of personal terminals, and they computed and communicated as a public good, subsidized by the federal and state government. They watched movies on their CRT displays, and they manipulated chemical bonds with the touch screens on their plasma displays. They were not consumers. They were users, yes, and in many cases they were authors, the creators of new PLATO programs. They were computing citizens.

The relentless emphasis on the legend of Engelbart, PARC, and Apple masks the existence of the thousands of people who crafted computing on personal terminals in Illinois during the 1960s and 1970s. The legend obscures the fact that there were once computing citizens who were not consumers. The mythology that centers on Silicon Valley also masks the important role that other regions played in creating our contemporary digital culture. Illinois mattered, as did the context in which CERL had been created. The personal plasma terminals of PLATO emerged from the Golden Age of Education. Students and teachers from the primary, secondary, and university levels all became citizens of PLATO's republic. Their use of PLATO for research, activism, communication, and recreation is analyzed in Chapter 7.

7

PLATO's Republic (or, the Other ARPANET)

Valarie Lamont urged her University of Illinois community to participate in the first-ever Earth Day activities during the spring of 1970. Lamont studied political science as a graduate student at Urbana-Champaign, and she focused her environmental activism on a small stream, the Boneyard Creek, that flowed through the two neighboring cities. With the help of a student group, the Concerned Engineers for the Restoration of the Boneyard, Lamont researched the history, flooding, and pollution of the creek, and she investigated solutions for the flooding and the pollution. Then, she deployed her research to persuade her peers to action by creating a computer program.

Lamont used the distinctive features of the university's PLATO (Programmed Logic for Automatic Teaching Operations) terminals to advance her cause. The PLATO project had begun a decade earlier as an exploration of computing for education, but by 1970, the system encompassed over seventy terminals on campus and at other locations around Illinois. Each personal terminal included a video screen capable of displaying photographs and films, as well as a keyset with which the user could communicate with the central mainframe computer that powered the system. Lamont incorporated photographs of a pristine creek and of creek pollution into her PLATO program, and she encouraged her users to ask questions and provide suggestions to her via the program. For Lamont and her users, the personal terminals of the PLATO network offered a new form of political participation.

Lamont's repurposing of PLATO, from educational system to activist platform, epitomized how users crafted personalized computing as the system became more popular during the 1960s. When the engineer Donald Bitzer imagined the system, he was working in a recently declassified but still military-sponsored Cold War laboratory. Yet the growth and success of his project required the cooperation of scholars and practitioners in the humanities, social sciences, and education, and he energetically recruited them. Over the course of the 1960s, not only did the number of PLATO student users increase dramatically, but so did the number of PLATO authors, individuals who produced their own programs on the system. Lamont was one such author. The constantly growing group of authors enthusiastically employed PLATO for educational, research, and recreational goals.

During the 1970s, users welcomed the new features of the expansive PLATO plasma-screen network, especially its communications features. By 1975, the 950 terminals on the nationwide PLATO network enabled "on-line" communication in the form of bulletin boards, instant messages, and electronic mail. This rich social network was partially funded by the Advanced Research Projects Agency (ARPA). ARPA is now known for supporting a completely different computing network, the ARPANET, which became a foundation of the Internet. Privileging the ARPANET-becomes-Internet history has caused us to overlook the other forms of networked sociability that emerged during and since the 1970s.

This chapter examines how PLATO users made themselves into computing citizens. I consider the activities by which they transformed a system that began as a teaching machine into a widespread digital community. In doing so, they changed PLATO from a limited Cold War network into a vibrant computing community. These citizens developed practices that are now integral to our modern digital experiences. PLATO people swapped jokes and stories on their online network, and they reveled in this new sociability. At the same time, they struggled with security, censorship, and harassment, and their interactions revealed a gendered digital divide.

The first section analyzes how Lamont and her graduate student peer Stuart Umpleby reimagined PLATO as a political tool. Lamont's and Umpleby's work illustrates how PLATO authors pursued their own goals on the system, and how new communities coalesced around PLATO usage. The next section compares PLATO's network with ARPANET. ARPA's support of the PLATO project stimulated its growth as a communications network, one whose users eagerly forged new forms of social computing. The final section, based on four years' worth of PLATO notes (the system's online bulletin board system), scrutinizes the diverse digital interactions of the citizens of PLATO's republic.

Alternative Futures

Maryann Bitzer's work to train nursing students using PLATO highlighted the problem of how to create new PLATO lessons, and it offered a solution. In her thesis acknowledgments, Maryann thanked Donald for programming the logic for her nursing experiment. During PLATO's first years, the small team had to program any new lessons themselves. They had to create the software to provide content for students. As interest in PLATO grew, the team wanted to empower others to produce their own lessons and programs, and the simulated laboratory offered them a way to accomplish that. Created during the fall of 1962, the PLATO lab, separate from the main or help sequence, enabled students like Maryann's nurses to run their own simulated experiments.[1] A few months later, the team realized that the lab offered "the possibility of a very general master program for PLATO which [would] allow non-computer trained persons to write their own 'teachers' for whatever kinds of lessons they desire."[2] During the next year, the team introduced a series of lessons to teach potential program designers—"authors, not students"—about PLATO "without involving them in all the technical detail."[3]

Maryann helped usher in the era of PLATO authors. The authors created new PLATO lessons, shared ideas with each other, and formed communities. In short, they produced new knowledge. They were not consumers, and they were not end users. Their ability to produce new programs, collaborate, and communicate on the system opened the door for a republic of PLATO citizens. Certainly, a hierarchy emerged. The people who created lessons were known as authors, and they were distinguished from the (generally more computer-savvy) people who programmed the system on the mainframe, as well as from the (generally less computer-savvy) students who used the terminals for lessons. In 1967 the introduction of a new programming language, TUTOR, simplified the process of creating new lessons.[4] This opening of the PLATO system to authors, together with a growing communications network of increasingly user-friendly terminals, transformed it from one of passive learners to active—and activist—citizens.

A few months after PLATO author training began, two project members added a mere twenty words of code that ultimately transformed the system. They created a subroutine called CONNECT, used for communication between terminals. The team celebrated the fact that such a small change to the system yielded so many new computing possibilities. As the team members explained, "the capability of communication among stations" opened up "a new dimension of PLATO research," including gaming, simulation, and group experiments.[5] CONNECT marked the beginning of PLATO as a personal communication and recreation system. Early in 1965, users gained the option to provide comments in their lessons, another communications option. Users could enter a comment at any point in the lesson, addressing the instructional material, the keyset operation, their fondness for the university—whatever they felt like imparting.[6]

Illinois graduate student and PLATO author Stuart Umpleby envisioned the system as a corrective for the pervasive upheaval in American society. Umpleby had earned bachelor's degrees in engineering and political science at the University of Illinois in 1967; the following year, Martin Luther King Jr. and Robert Kennedy were assassinated within months of each other. Teach-ins at campuses around the country protesting the Vietnam War marked Umpleby's undergraduate years, and the counterculture, Black Power, Red Power, and Chicano Power had erupted during that time.[7] As a graduate student in political science and communications, he sought to alleviate "the present domestic instability and sense of foreboding in the United States."[8]

Working with Charles Osgood enabled Umpleby to address those concerns. Osgood, a psychologist, had begun using PLATO for his experiments in 1965.[9] He first studied interpersonal norms using animated films on PLATO. Osgood and Umpleby then worked together, beginning in 1967, on possibilities for the future. They wanted to create a set of tools through which individuals could explore scenarios for the year 2000.[10] Individuals could rate the desirability of a future outcome, such as reducing global pollution, as well as the likelihood of that future outcome, and they could observe how ranking one outcome as highly desirable might reduce the likelihood of another outcome. Osgood and Umpleby thought it was valuable and eye-opening for individuals to confront such choices.

They initially framed their project, called DELPHI (like the site of the ancient Greek oracle), in the language of futures research, a burgeoning field. Futures research entailed planning for the future (including urban planning), developing methods of forecasting, and involving the public in both planning and forecasting.[11] GE had established TEMPO (Technical Management Planning Organization) for this purpose, and the Air Force's already well-known RAND Corporation had been founded to coordinate long-range planning with government research and development decisions.[12]

As he worked intensively with PLATO on the DELPHI project, Umpleby reimagined the possibilities and the purpose of the system. He realized that PLATO was not just a teaching computer, limited to the realm of education. It was a "mass communications system with feedback."[13] In other words, thousands of future PLATO users could receive text, images, and audio, and they could provide text-based responses to whatever they had just read, seen, or heard. Umpleby compared this medium to radio, television, mass rallies, town meetings, and even Congress. For example, Umpleby argued that "television and radio are evanescent. The viewer or listener has no opportunity to go back and examine the logical argument or to check a point he missed while his mind was diverted by an earlier remark."[14] Umpleby's comparisons to mass rallies, town meetings, and Congress were not coincidental. As a political scientist interested in the future, Umpleby sought methods for expanding democracy.

Umpleby proposed "citizen sampling simulations," whereby citizens used PLATO to learn about a particular local or national policy issue and communicate their preferences about outcomes to policymakers and legislators.[15] He described another PLATO game, POLIS, to show how this might work. In POLIS, the user assumed the role of a small-town police chief whose primary responsibility was upholding free speech. The police chief had to learn about and balance the perspectives of a "militant speaker, a conservative town council . . . and a crowd whose mood chang[ed] during the course of the 'game.'"[16] POLIS allowed the person in the role of police chief to receive advice from a "city attorney" in the game before deciding on a course of action that was "defensible by constitutional standards."[17] Playing POLIS forced the user to evaluate different sides of upholding the right to free speech. During the widespread protests and conservative backlash of the late 1960s, Umpleby's examples resonated with his peers.

Lamont shared Umpleby's interest in the digital citizenship opportunities afforded by PLATO, and she authored her program during the spring of 1970, in the months immediately preceding and following the first Earth Day.[18] Lamont's program, *Boneyard Creek*, presented the biography of a stream that ran through the communities of Champaign and Urbana. It described the stream's history, flooding, and pollution, and offered several solutions to the environmental issues. As Lamont wrote the program, she consulted a report that had just been written by a group of Illinois students who called themselves the Concerned Engineers for the Restoration of the Boneyard, or CERB.[19] Lamont transformed the CERB report into a compelling narrative for her users. Users navigated the *Boneyard Creek* program to gather information about unfamiliar terms, view photographs of the creek and its pollution on their screens, and provide their comments and opinions to Lamont at any point along the way, using the keyboards located below their screens.[20]

Lamont's topic stemmed directly from the community's existing concern for environmental issues, and she wanted to evaluate PLATO's potential to stimulate "citizen participation in community planning."[21]

Lamont situated her work firmly within the social movements of the long 1960s. Citizens wanted more opportunities to participate in policy formulation, and they wanted new methods of participation, she argued. She sought an issue that was important to residents in both Urbana and Champaign, that entailed medium- or long-range planning, that was familiar to most people, and that was "relatively non-political but somewhat controversial."[22] During the fall of 1969, while Lamont considered potential issues, the Concerned Engineers called attention to their Boneyard Creek cause with a well-publicized cleanup day, and they searched for ways to propagate their environmental mission after they graduated. CERB's and Lamont's activism promised symbiosis.

Lamont's Boneyard program, first shared with the Urbana-Champaign public in June 1970, represented a personal computing approach to—and extension of—the April 1970 Earth Day teach-ins. The Wisconsin senator Gaylord Nelson had called for a nationwide teach-in: a series of lectures, discussions, and demonstrations around the United States to raise awareness and generate additional support for local and national environmental concerns.[23] Lamont described each iteration of the experiment—that is, each gathering of individuals to use the program—as a "demonstration."[24] About nine people met for each demonstration, and their personal terminals provided the educational forum. Lamont emphasized the interactivity of PLATO, especially compared with such traditional modes of information delivery as newspapers, radio, and television. Lamont's users could seek out definitions for unfamiliar terms, they could RETURN to previous slides to review information, and they could provide feedback about the program and its contents at every step of the way. Such comments included "can laws be passed which would regular flow of sewage into boneyard in future" and "the cost of landscaping alone would be less than sheetpiling, but what about the added cost of eliminating pollution?"[25]

At the end of the program, Lamont urged her users to additional action. The options that Lamont suggested for further community participation included writing letters to government officials, purchasing property (and then donating it for recreational areas), calling for rezoning, undertaking cleanup and landscaping projects, and, of

PLATO terminals at Parkland College, Champaign, Illinois, circa 1969. Photo courtesy of the University of Illinois at Urbana-Champaign Archives, image 949.tif.

course, inviting other friends and colleagues to use the PLATO *Boneyard Creek* program.[26] Lamont embraced a new technology as a form of environmental activism. She created a program that showcased an environmental issue, offered possible solutions, and called people to action. Lamont's approach of education as a means to activism mirrored the efforts of other social movements of the time, especially the civil rights movement.[27]

The experience of using the *Boneyard Creek* program was both personal and social. Users sat at their own workstations, with their own keysets and screens. The screen greeted users with the message, "Hi. My name is PLATO. I am the computer. I talk to you by writing on this screen. You talk to me by pushing the keys on the keyset in front of you."[28] Users proceeded through the slides at their own pace, and they could request and obtain additional information about selected topics in the program. The program invited users to provide observations and suggestions about PLATO, the program, or the creek throughout, and with directed questions at the end. Each user experienced individualized, interactive computing but shared the experience with the ten or so other people in the PLATO workroom at the same time. Users who had a question or encountered a problem could request help from the assistant on hand, perhaps Lamont herself. A user could speak to a neighbor in a nearby cubicle, remarking on the pollution he or she had seen firsthand at the creek. They could decide, at the conclusion of the program, that they would meet in the coming week to write to their local city councilors. The user simultaneously experienced the privacy of his or her own terminal and the public space of the PLATO laboratory, a space at once both personal and communal. Although the user viewed the program individually, he or she was called to action as a citizen, a member of the local democracy.

Lamont's appropriation of PLATO to encourage environmental activism, and Umpleby's recognition of its potential as a two-way mass communications medium signaled the changes under way for the system. Neither Lamont nor Umpleby was a passive consumer; they were not just users. Both had authored programs, and both had imagined new forms of digital political participation. Moreover, they belonged and contributed to a larger community dedicated to exploring PLATO's potential. The Alternative Futures Project explored the relationship among citizenship, technology, and civic participation.[29]

Lamont, Umpleby, and their Alternative Futures colleagues investigated the pressing issues of the day: legalized abortion, nuclear war, animals as organ donors, human cloning, genetic manipulation, legalized marijuana, a national data bank, population planning, and a World Aid program, among others.[30] Similar PLATO communities arose, for example, among the authors developing lessons in particular subject areas and among the authors creating games.[31] Another community rapidly recognized the powerful communications features of the PLATO plasma-screen network when they began using it during 1972–1973. That year, the PLATO team worked with a new group of authors, users, and sites through a project sponsored in part by ARPA.

The Other ARPANET

ARPA awarded the PLATO group a contract for the exploration of "computer-based education for a volunteer armed service personnel program" to run from 1972 through 1975.[32] Just as the Soviet launch of the satellite *Sputnik* in 1957 spurred widespread reforms in American public education during the 1960s, so, too, did it propel the military and President Dwight Eisenhower, a five-star general who had commanded the Allies in Europe during World War II, to take action. American leaders feared that the Soviet Union would gain the upper hand in the technologies and weapons race of the Cold War. They created ARPA within the Department of Defense to ensure that the United States led the world in scientific and technological development, never to be ambushed again as they had with *Sputnik.*[33] The agency cultivated and funded projects with potentially revolutionary benefits for both the military and civilians. The agency's decision to evaluate PLATO for training military volunteers reflected a continuation of the project's long-standing relationship with the armed forces. After all, when the project originated at the Coordinated Science Laboratory, Bitzer had emphasized its value for military education.

From 1972 through 1975, a network of ARPA-sponsored PLATO terminals spread across the United States. The PLATO team installed and operated PLATO plasma terminals at eleven military sites across the United States.[34] By January 1974, there were 50 PLATO terminals across these eleven sites, with plans for another fifty terminals within six months. The PLATO terminals at each of those far-flung locations were connected to the main PLATO Control Data computer on the University of Illinois campus.[35] Thus, the terminals were also connected to each other through the central PLATO computer. By extension, the ARPA-project terminals were connected with all the other PLATO plasma terminals in Urbana-Champaign and elsewhere. They formed a network of ARPA-sponsored (and NSF-sponsored) PLATO terminals through which people could communicate with each other.

This PLATO network, partially funded by ARPA, existed contemporaneously with the ARPANET, widely recognized today as the foundation of the Internet. ARPA had funded the Cambridge, Massachusetts–based company of Bolt, Beranek, and Newman (BBN) to build the ARPANET, that is, to network different computers at different locations with each other. The first ARPANET transmission, in October 1969, traveled from the University of California at Los Angeles through BBN's purpose-built interface computers in Cambridge to the Stanford Research Institute in Menlo Park, California.[36] By 1972, ARPANET had grown to thirty-seven nodes. However, as one historian of the Internet has argued, the early ARPANET was not a particularly hospitable—or useful—place for the people who accessed it.[37] During its first few years, ARPANET users experienced unreliable connections. They were often stymied by the incompatibility of the different makes and models of computers connected to the ARPANET, which impeded sharing software and data. A 1972 report on ARPANET written by an outside consultant observed that "the network user, new and established, is probably the most neglected element within the present development atmosphere."[38]

Because ARPANET had been created to connect and share the valuable resources of computers, most people did not immediately consider the possibilities of sharing human resources. Rather, users valued ARPANET only after they created an unplanned use for it: electronic mail. ARPANET's "smash hit" was its role as a communications medium.[39] E-mail began to circulate on ARPANET during 1972–1973, but the first widely popular e-mail reader was not created until 1975. These were exactly the same years during which ARPA funded the PLATO network.

While ARPANET users only gradually came to value their network for communications, the ARPA users of the PLATO network immediately recognized its value as a communications medium. In fact, the first PLATO team report to ARPA on the project is striking in its devotion to the various types of communication that were possible and popular on the PLATO network. The report authors highlighted the communications features as novel, and a significant advantage, very early in the report. They noted the wide geographic distribution of the PLATO plasma terminals, and they asserted, "As a result of this geographical dispersion and of the urgent need for rapid communication between users at different sites, inter-terminal communication has become, not simply a possibility, but rather a requirement."[40]

The PLATO network functioned with such versatility as a communications medium that the PLATO-ARPA team created descriptive categories for the different kinds of communication that the plasma-terminal network enabled. They differentiated communication along the axes of directionality, focus, and immediacy. With "directionality," they described a one-way flow of information compared with a two-way flow of information. For example, a user posting an announcement that did not require a response (sending out information) contrasted with a user asking a question that did require an answer (sending and receiving information). The authors used "focus" to categorize how many people would be interested in a particular communication; that is, was the communication for everyone on the system, for people working on a particular course, or for just one individual. "Immediacy" described desired response time: did the communication require a response within minutes, hours, or days, or not at all?[41]

The PLATO-ARPA team analyzed the network's range of communication options, which included features we would now recognize as instant messaging, screen sharing, digital message boards, and e-mail. If two people wanted to communicate directly with each other, and both people were present at their terminals, they could use the "talk" feature (two-way, high immediacy). These two users then shared a written "conversation" on two lines near the bottom of their respective terminal screens, with the words appearing "immediately." If several people were present at their terminals, and they all wanted to address a topic together, they could use "talkomatic" to share on-screen text-based conversation via an "on-line conference." The PLATO-ARPA team also created an extension of the "talk" feature, by which a user could request assistance from a PLATO consultant while the user was writing a TUTOR program. The user and the PLATO consultant, possibly hundreds of miles away from each other, shared a "talk" conversation on their PLATO terminals, but the consultant also saw and monitored on his or her screen part or all of the user's screen (depending on how much access the user provided). This screen sharing allowed the consultant to see where the user might be running into problems, and to advise the user on potential solutions—all via the PLATO network. The PLATO-ARPA network also offered various "notes" files (such as "system notes" and "consult notes") in which users could report problems and make suggestions for future developments. Authors on the network, those users who created their own TUTOR programs, could read and contribute to multiple notes files. The PLATO-ARPA staff characterized the notes files as a "historical record of system development." The PLATO team also created "bulletin board" features, ranging from "notices that new features are present" to "an informed 'newspaper' available on-line." In 1974 the PLATO-ARPA team implemented a feature that is perhaps most notable to us today. Called "itc," it enabled one user (Joe) to send a message directly to another user (Jane), and Joe's message would appear on Jane's terminal whenever she next logged onto the system. Jane could also respond directly to Joe's message at her convenience. The PLATO-ARPA team commented on this "more sophisticated routine" that "messages for a particular individual or answers to his own questions are automatically presented to him at the next time he signs into the program."[42] A year later, when the PLATO team produced an internal report on plasma-terminal communications, they called this feature "electronic mail."[43]

The PLATO network was striking both for its widespread use for communications and for its range of communications options, a combination that engendered a rich collective. ARPANET users struggled with incompatibility, but PLATO network users benefited from interchangeable terminals all connected to one central—and centrally maintained—computer. All of the PLATO terminals were compatible by design. ARPANET users experienced network reliability challenges because their network was young, their access was limited, and their hardware sometimes failed.[44] But by 1972, PLATO users enjoyed the fruits of hardware, software, and maintenance expertise developed over the past twelve years.[45] The PLATO-ARPA team even collected data on how frequently users employed the system's various communications options. During a typical week in 1973, while the plasma-screen system was still expanding (at the time, about 380 terminals), over 500 entries were posted to the various notes files, and 175 messages were sent via the "itc" program.[46] The project leaders observed that this digital communication "resulted in a remarkable sense of community."[47]

In contrast with the neglected new users of ARPANET, PLATO's new users enjoyed multifarious human and computer resources to help them navigate the system. The plasma-terminal network users could access lessons called "help" and "aids" to gain information on topics ranging from the touch panel and the keyset to log-in procedures. Other lessons addressed the TUTOR programming language, a schedule of terminal usage at each terminal location, how to use a basic statistics package, how to use PLATO to poll student opinions, and how to troubleshoot a terminal. If, on the other hand, a user simply wanted to interact with a person, he or she could use the "talk" feature or send a message.

Step-by-step instructions for using the "talk" option were provided on the first page of the PLATO IV user's manual, which underscored the centrality of "talk"—and user communication—to the network. The manual assured readers that "even if you cannot contact the individual on-line, there is a section of 'notes' where a personal note can be left. . . . Your note will be stored and saved until it is seen by the addressee."[48] Indeed, throughout the manual, users were encouraged to "talk to" a particular person for a certain question or issue. The manual provided the first and last names and "course" for each contact person, which served as a person's address on the system.[49]

By 1975, the PLATO network, the other ARPANET, connected users at 145 locations around the United States, including 26 locations on the University of Illinois campus; a terminal at the University of Stockholm in Sweden marked PLATO's international expansion. Approximately 950 terminals spanned North America from California to Massachusetts and from upstate New York to Florida. During 1974 and 1975, total terminal usage—by both students and authors (those creating programs)—ranged from a low of 40,000 terminal hours per month in December 1974 (during the holidays and semester break) to a high of 120,000 terminal hours per month for October 1975. Thousands of users charted over one million hours of screen time during the first eleven months of 1975. The plasma screens of this other ARPANET glowed with their activity.[50]

Notes from an Early Computing Community

The military personnel who participated in the ARPA-sponsored project to evaluate PLATO were just some of the hundreds of authors who contributed to the PLATO online community, one that was documented through the preservation of the system notes files. Every few weeks from October 30, 1972, through June 13, 1976, one of the system programmers printed out a hard copy of the online notes files.[51] Variously called "lesson notes," "general interest notes," and "public notes," the messages posted to this online bulletin board were supposed to be of general interest to most authors and programmers who were using PLATO. These general interest notes were differentiated from "help" notes (also known as "consult" or "consulting" notes), which were intended to advise authors who wanted assistance with creating and maintaining lessons in the TUTOR programming language. Furthermore, PLATO IV also featured something called "pad," wherein authors could post personal notes and advertisements. All of these files were public in the sense that all authors could read and respond to whatever was posted, but only the notes files were saved over time. The "help" and "pad" files were recycled every few weeks.[52] Despite the fact that the notes focused primarily on system or TUTOR errors, or other issues of widespread interest, the nearly four years' of exchanges evinced the now-familiar characteristics of online communities. The users were primarily system programmers and administrators, consultants who worked with the PLATO Services Organization, and hundreds of authors around the United States.

The authors had a love-hate relationship with the system programmers who supported them: the programmers solicited and considered author feedback to a remarkable extent, but the programmers also wielded tremendous power over the authors in sometimes inconsiderate ways. During the 1972–1973 academic year, the programmers and authors alike prepared for the official PLATO IV release in

September 1973 while they tested a gradual rollout of the new plasma-screen stations and related equipment. The programmers reported their attempts to design a new character set for the new screens, and programmer Andrew requested author feedback. Two authors agreed that the proposed characters were "awfully distorted," and they did not like them.[53] Deborah and Larry mentioned character sets that had been designed by Bell Labs and the Air Force, and a few weeks later, the programmers proposed another character set for author consideration.[54] The discussion about character sets for the plasma screens also included author input about characters to be "wired in to the character memories" or to be deleted.[55] One user requested forty-five-degree-angle arrows; another asked for a cents sign. Two requested keys to jump to the beginning or the end of a block of text to facilitate their own navigation.[56] The programmers considered and weighed in on these and similar requests.[57] The author Kevin and the programmer Jeffrey bandied back and forth about the possibility of a "scroll type command" for advancing through text on the screen.[58]

Here, the programmer (Jeffrey) even sought the advice of the author (Kevin). Jeffrey mused, "How do you people about the world operate without print-outs? Do you take polaroid pictures of your lessons. . . . Several of us at home-base would like to know your solutions . . . to tell others. In the long run, we expect there to be terminal printing devices." Jeffrey acknowledged that authors who were based in Champaign-Urbana could request printouts of their PLATO lessons for convenient reviewing, editing, and bug fixing. But Kevin was one of the many remote authors, and Jeffrey hoped to benefit from the adaptive PLATO skills that the distance necessitated.[59]

When one PLATO consultant sought the community's feedback on new "author aid lessons," vigorous discussion ensued. Some thought the proposal was "terrific" and "great," while another "thoroughly disagree[d]."[60] The extraordinary give-and-take between the authors and the programmers prompted staffer Kenneth to declare, "Users have more influence over the development of this system than any other I've heard of."[61]

Indeed, the PLATO programmers and consultants diligently responded to most queries and comments on the notes, although with dramatically varying degrees of politeness and consideration. To be fair, the programmers endured a heavy workload, especially during the 1972–1973 year when they prepared to officially roll out the PLATO IV network. In March 1973, when author Carol highlighted the need to fine-tune the "answer-judging function" on the system, programmer Jeffrey reported, "I agree, but my boss has lots for me to do."[62] Jeffrey, at least, demonstrated sympathy in that particular response. Some of the other programmers were less than kind. A few months earlier, in December 1972, when a PLATO citizen offered suggestions on the proposed character sets, programmer Andrew publicly mocked the citizen by posting, "If anyone other than [citizen] can make any sense out of his comments, please tell me."[63]

Andrew cultivated a role as both sarcastic moderator and PLATO policeman, frequently posting admonishing comments such as, "Repeating an old, old, refrain: notes signed with initials only will not only be ignored but will be deleted."[64] The PLATO programmer Anthony referred to a user as "some twit," while Andrew often posted condescending messages that the authors were not making any sense.[65] Indeed, authors lamented the "smart-alecr [*sic*] answers from the 'old pros' among systems people."[66] For their part, the programmers decried the "snottiest notes" posted by "new authors," and Andrew stood up for all of them when he asserted, "Believe it or not, the people working on the system are also human, and often find it difficult not to respond in a fashion similar to that employed by some of our users."[67] Nonetheless, the programmers offered apologies when they recognized the error of their posting ways—or their actions on the system. When the programmers changed a keycode that made multiple courses "inoperable," author Joseph chastised them: "One system programmer's whim can now cost dozens of people hundreds of hours."[68] The programmer Jeffrey sheepishly responded, "We made a mistake. . . . All we can say is S<RRY [*sic*]. We shall try not to do this type of mistake again."[69] For the most part, the programmers aspired to balance the technical needs of the system, such as providing simultaneous service to hundreds of terminals, with the requests of their users.

Some of the exchanges occurred between authors and programmers, but many others demonstrated collaboration among authors, and they hinted at the extensive community who read the notes without frequently posting. One user remarked, "To the author of the nine coins game: it allows two players to choose the same coin," and shortly thereafter the game author posted a thank-you in response.[70] The comment was offered in a spirt of helpfulness, and the response demonstrated that authors eagerly read new notes. Enough of these one-off exchanges occurred over three-plus years to corroborate a high level of readership and engagement. In January 1973, Perry inquired whether anyone had a PLATO-appropriate map of the United States, and just two days later, another PLATO citizen confirmed that he did. Meanwhile, Denise informed Perry that "[another PLATO person] (who is on vacation I think) is working on a map of Europe. She might also have U.S. I would like to have U.S. also. Maybe we can get together."[71] That response epitomized the communal and cooperative nature of PLATO's republic, and the ways in which authors shared information and resources. Similarly, when one PLATO person wanted a "command that draws a broken line," another responded that she had one, and she offered her room number and telephone extension.[72] Authors even stood up to programmers on behalf of other authors. When system developer Philip castigated two authors about their posts, fellow author Joseph defended their messages as logical and useful. Joseph then went a step further to suggest alternatives to Philip that would be helpful to those authors and their peers.[73]

Considering what we now know about etiquette in online communities, it is not at all surprising that the users also policed and criticized each other. After Joseph commented on a proposed TUTOR change, another user endorsed Joseph's suggestion, but misspelled Joseph's name. Joseph shot right back, "!!![User]. If I can spell [your complicated name] you can spell [my less complicated name]!!!"[74] As Joseph suggested, attention to manners and the spelling of colleagues' names went a long way toward creating collegiality when interacting via screen. Some peer policing was intended to be helpful (although perhaps with a tone of admonishment), such as when one PLATO person addressed, "[Person A] or [Person B]: Probably due to your inexperience with TUTOR authoring, I believe you accidently deleted a note that I left earlier today. Please be careful."[75]

Other peer policing was decidedly caustic. In February 1973, Samuel posted a helpful note telling other authors how to rotate upside down any "homemade characters" they had created (like letters or numbers or atoms or flies). In response, "Bill Haywood" (likely an alias, given that Bill Haywood had been a prominent labor activist and socialist) posted a cruel reply: "Real good, [Samuel], but what use is it. This notes file should perhaps not be used to describe every esoteric routine somebody develops. The number of authors that wish an aid to turn RAM characters up side down is almost nill."[76] Haywood implied that Samuel's work was valuable to no one, and that Samuel had wasted fellow users' time—and valuable system space—by sharing this suggestion. Others came to Samuel's defense, and PLATO services consultant Larry finally stepped in to apologize for Haywood's comment and to clarify official policy: "Mr. Haywood's no doubt unintentionally abrasive evaluation certainly does not represent [C]ERL policy on the use of these notes. Anyone who feels that he has something of interest to other authors should feel free to mention it in -notes-."[77]

Despite a general spirit of cooperation, security and system stability remained a concern for PLATO users. Authors reported break-ins on their TUTOR lessons, and they sought to protect their work. System administrators warned of security breaches and tried to implement backup options, but they also sought to distance themselves from responsibility for security. The programmers and the authors each thought the other group was primarily responsible for ensuring the security of author lessons. In November 1972, one PLATO person asked, "Is there any way to protect work being done in edit mode when the system goes down? I lost two blocks in the past two days."[78] System administrator Andrew promptly advised, "If your [*sic*] are worried about system stability, you should back out of the block you are working on after doing enough work that you do not want to lose it. Your work to that point will then be (more or less) securely preserved on the disk. You can then return to work knowing that much is saved."[79]

Andrew's parenthetical note that work would be "more or less" securely preserved indicated one of the major problems with PLATO IV at that point. It was susceptible to malicious intrusion. On December 10, another PLATO person was "reading in" (or saving) a block of work when it was "zapped," and he saw the message: "All right. You scrawny, pale system programmers . . . I want to see you up on your feet exercising!!!"[80] Two days later, Joseph reported the same issue: "Add one more victim to the list of those whose blocks get zapped. . . . That's four *?!:$$* hours down the drain."[81] Joseph's comment conveyed that there were several victims of the security breach, and a question posted on December 12 confirmed it. Deborah identified the security problem as the "file stomper" and asked: "To systems people—what is kept in the way of backups? I understand that there is a disk backup run about once a week. Is there anything else also? Since the file stomper is still going strong, I'd like to know so I can best plan for his attack."[82] Deborah recognized that the team was not yet able to stop the attacks, and she sought advice on how to protect her files.

Here, the notes reveal some of the early vulnerabilities with PLATO IV, and problems from a user's perspective. Complex TUTOR programs were time-consuming to create, and PLATO did not autosave them. Rather, the author had to "back out" and "read in" a block of work to save it before continuing. Even then, the "read in" could be interrupted by the file stomper. In response, the systems people suggested printouts once the TUTOR program had been "read in," but users reported not receiving the printouts they requested. On top of all that, there were no printouts on weekends, when many authors developed their programs.[83]

In 1973, the PLATO people faced new security challenges. On January 23, Barbara warned "ALL AUTHORS" that "Several programs have been written that find important information left in the computer about each of your lessons," including the file name, the change code, and the inspect code.[84] Barbara reported that such information had already been misused at least once, and she urged authors to change their security codes frequently. Later that day, the user who identified himself as Bill Haywood observed that the idea of security on any computer system was an unattainable goal. Haywood declared:

> I bet the new "leakproof" version of PLATO stays that way about two weeks. Anyone with a basic knowledge of computer systems knows "system security" is a figment of the system program-

> mer's imagination. Any user who is determined and smart enough can outwit system security. The safety of your lesson is dependent on the good will of those who try such things.[85]

Two of the systems programmers responded to Haywood with varying degrees of defensiveness. Anthony posted: "Pardon me. . . . I didn't realize I had claimed to produce a 'leakproof' system." Meanwhile, Jeffrey weighed in with the observation that he still considered PLATO IV to be experimental and not very secure. He claimed that he and the other systems people had "not placed much emphasis on security except to prevent system crashes," but once they did, they expected PLATO IV to be "totally secure." With his statement, Jeffrey delayed dealing with the thorny security issues until a later date. A few weeks later, Larry issued another "WARNING TO ALL AUTHORS," informing them of a new program that mimicked the PLATO sign-in screens and procedures and thereby captured access and change codes.[86] Security problems pestered PLATO users intermittently for the next four years, as the system administrators created new levels of access that were then thwarted by "file stompers."[87]

On the PLATO network, the issues of security and identity were intertwined. Authors accessed their programs using their names and security codes, and programmers frequently exhorted them to change those codes. (Many of us working with networks nearly fifty years later are quite familiar with these admonishments to select new passwords often, and to make the passwords very clever and "strong.") Exact program or lesson names were supposed to offer a measure of security to authors since those names were not widely publicized. In January 1973, an administrator warned the authors via notes that all users were not supposed to know the name of a lesson unless the author had revealed that information on a person-by-person basis. Thus, if a user was interested in a particular subject area, program, or lesson, that user needed to search the PLATO "catalog" program to find an author's name and contact information.[88] A few weeks later, concerns over security and identity verification came to a head when author Randy discovered that someone had stolen his PLATO identity. He reported:

> Someone tell me what is going on! As of 1/20/73 my lesson bones was intact. I returned from semester break today and found some strange material in bones. My security code was changed and an inspect code placed on the lesson. Not only that, but whoever was responsible had the gall to ask to retrieve [Block A] of bones, and (the same person, I assume) sign a ridiculous note about spacewar with my name. . . . HELP![89]

Randy's complaint highlighted a recurring concern with computing and networking security: how did someone prove his identity? How did the PLATO community determine which actions were fraudulent and which were valid? In this case, PLATO staffer Kenneth posted to notes: "Will the real [Randy] come to see me."[90] Presumably this solution worked, because Randy lived and worked close to the PLATO hub in Urbana-Champaign; but as the network grew in size and geography, such problems became more challenging to resolve.

Indeed, the issue remained unsettled two years later, when Frank wanted to call the PLATO offices to change his security code. Denise responded, "Since presumably anyone could telephone and request a security code change, you will probably have to appear in person if you are in C-U [Champaign-Urbana]. . . . Any systems programmer can help you—if you convince him that your request is justified. If you are out of town, perhaps your course director is the one to ask."[91] Denise claimed that an in-person appearance was the only way to truly decide the question of identity and security, and for those cases in which an in-person appearance was impossible, then validation by the proxy of the course director—who presumably knew the author personally—might suffice.

The complex interplay of real-world and online identities rapidly manifested on the network, especially when it came to matters of age and gender. PLATO author Stewart Denenberg observed that "since the communications in games as well as in notesfiles are stripped of their stereotypic cues (sex, age, race, tone of voice, physical characteristics, smells, body language), the communication becomes truly egalitarian." Denenberg admired the PLATO system, and he presented an idealized view of it. He celebrated the ways in which

PLATO "enhanced and humanized" conversations and interactions. He fondly reported receiving assistance from another mystery user when he began playing the complex game *Empire*. Denenberg recalled that "after many minutes of valuable instruction, I was asked by my mentor, 'How old are you?' '38', I replied. 'YIIIKES', was the response. 'What's the problem?', I puzzled. 'I'm 12', replied the mentor. 'NO problem', said I—although it was very difficult for me to avoid talking 'down' and the 'mentor' from talking 'up' in the communications that followed."[92] Denenberg's status as a white man using the network mattered. He enjoyed the novelty of receiving guidance from a kid, but not everyone shared that sentiment.

Many of the authors attributed network security breaches to kids and decried them, but the authors also celebrated a system programmer who was a teenager. During the frequent "file stompings" of 1972–1973, authors found parts of their carefully programmed lessons replaced with "garbage" text, or, while programming, they were suddenly "jumped out" of their lessons in progress into a game like "racetrack" or checkers.[93] The user posting under the name "Bill Haywood" suggested that "reliable rumor has it that the program to reveal security codes was devised by a 13 year old kid." He then asked, with a touch of sarcasm, "PS. Why do you thin[k] the university figures out student bills, grades, etc. in the basement of Admin. Bldg." rather than in the computer lab?[94] The programmer Anthony, in response, scoffed, "Perhaps we should move PLATO to the basement of the Admin. Bldg" to make it more secure. But an anonymous user had the last laugh with the wisecrack, "Perhaps we should move all 13 year old kids to the basement of the administration bldg."[95] This response reflected the attitudes of more than one author; indeed, the association of teenagers with what we would now call "hacking" PLATO was already well established.[96] One author declared, "Someone should be trying to improve security weaknesses brought out by the antics of these UniHigh system crashers," referring to students at the University (of Illinois) High School.[97] Months later, PLATO employee Kenneth warned, "All Authors: Many author records have been pirated by young people who have no other way to get onto the system," and he (and other administrators) urged frequent changes in passwords as a measure of protection against such piracy.[98] Nonetheless, many of the authors frequently interacted with—and expressed gratitude toward—the helpful system programmer Roger, and in March 1974, Jeffrey called attention to the fact that Roger was only fourteen years old.[99] In fact, Roger also attended University High School, from which he graduated in 1976, before he earned his bachelor of science degree in electrical engineering from the University of Illinois in 1979. He programmed for PLATO from 1973 through 1979.[100]

Although Denenberg commented on the lack of "stereotypic cues," identities were not obscured; the network was decidedly male dominated and often hostile to women. Nearly all the system programmers and PLATO services consultants were men, and most of the authors who posted notes were men. References to gender stood out. Notes by female contributors (whether authors or consultants) were also striking because the responses the women received dramatically differed from those directed to male contributors. Certainly the men were occasionally rude to each other in their notes, but even in those cases, apologies were often exchanged; at other times, the men offered their "pleases" and "thank yous," and they backed down if another contributor called them on poor behavior.

One such exchange occurred in February 1974 when authors Peter and Samuel vented frustration at the system programmers for a recent change, in a note titled "arrogance." After some back-and-forth, Peter offered regret for the "vehemence" of his comments. But then consultant Denise weighed in. She called out the "abusive notes" of five users by name, and four of them (possibly all five) were men, including Joseph, Samuel, and Peter. The responses to Denise were notable in their number, the hostility of their tone, and the personal nature of the comments, such as the one from Joseph: "Sorry you're so bitter." The offending authors backed down and even apologized when the male programmers and consultants (Andrew, Philip) stood up for themselves and offered another explanation of the situation, but no such apologies were tendered to Denise. However, one author did intercede on Denise's behalf. He observed, "I have been around for 20 months and this is the first time I've seen Denise lose her cool (re: response to note 179). Many users lose it several times a week. I hope this comment adds some perspective to the discussion."[101] Perhaps some of the people involved in this controversy also engaged in offline conversation—and apologies—but the on-screen and on-line comments clearly displayed more animosity toward Denise than any of the men.

Less than a week after the flame war involving Denise, three men launched another series of personal complaints against another woman consultant, Catherine. The three men (or young men) complained that they were doing perfectly valid work as authors when Catherine "began shouting at us" and "threatened the destruction of every game on the system." Subsequent notes revealed that those three men were actually writing a game (rather than the work they had been assigned), and their activity may have been disturbing other authors and students. Yet the three of them never apologized for accusing Catherine of "harassment" and "threats." Another system staff member even sought to distance himself from Catherine when he asserted, "We are not a homogeneous bunch of ogres," thereby implying that Catherine *was* an ogre.[102] Although some may argue that these incidents were minor and that too much may be read into them here, they did stand out for the differences in tone and overall trajectory of the conversation when compared against four years' worth of such notes.

The other women who regularly posted to notes received—and complained about—harassment, inappropriate comments, and patronizing attitudes. In November 1974, Teresa bemoaned this harassment in a note she called "crank call." Teresa exclaimed, "ARRRRRRGH! I am getting VERY tired of having people call me on 'talk' and ask what I am doing, if I am female, if my name is really '[Teresa],' if I know of any short games and other nonsense ad infinitum." The responses to Teresa's comment demonstrated the casual misogyny of the PLATO network. One user identifying himself with a typically masculine name suggested that Teresa remove her name from the "users" list, even though Teresa had clearly explained in her original post that she wanted to remain in the talk users list so that her collaborators could easily reach her. Another PLATO man at the University of Arizona suggested that Teresa develop "a sense of humor" about it, and then added the condescending comment, "If you are really bothered by others who are just being friendly over an impersonal machine I truely [*sic*] feel sorry for you." A third man echoed the Arizonan's comment, also patronizing Teresa: "I really feel sorry for people who cant [*sic*] say 'get lost' when they are up to their ears in work."[103]

These individuals, with markedly masculine names, had probably not experienced the frequency of crank calling that Teresa received, and almost certainly had not received the unwanted and inappropriate sexual attention. However, instead of offering sympathy to Teresa or helpful approaches to her problem, they criticized her complaint. They blamed the victim. Another woman, Sharon, chimed in and verbalized this clear difference:

> Taking your name off the user list is not a (excuse me, please) viable alternative, especially for consultants and people like [Teresa] who frequently get "talks" from people who need information. People who use last names or men's first names don't realize that some of us get "talks" asking us if we are REALLY working, if we are married, what we are doing friday night (or anytime), and so on. People who use the talk option that way should just QUIT, and not put the burden of dealing with it on me![104]

Sharon noted the different treatment that she and Teresa received by using their first names, and by the assumption that they were women. Moreover, she emphasized that the solution was that the people who were doing the harassing should stop. It was not her responsibility (nor any other woman's) to deal with this misuse of the system. Sharon, like Teresa, had also received patronizing comments in response to her notes. Sharon had been told by the system programmers (presumably in response to a query about TUTOR or the system) that "A little thought goes a long way." When she posted another question, and shortly thereafter posted that she "got [her] answer," the system programmers responded "We all love you [Sharon]."[105] One could read that response as enthusiasm or support; perhaps it was gentle ribbing. But in the context of having been advised that "a little thought goes a long way," the message "We all love you" seemed rife with condescension and innuendo.[106]

The discrimination against women on the network mirrored women's experiences on campus. During a notes exchange about the system being "over-subscribed" and the fact that "the game playing has gotten very loud," Peter observed that "some female students are very reluctant to kick someone out."[107] Although those women had priority on the system as students who were completing PLATO coursework, they had a difficult time—and possibly were harassed or mocked—when they asked the gamers, most of whom were young men, to leave. The harassment, on at least two occasions, turned physical. Between Thanksgiving and Christmas 1974, the CERL PLATO classroom and the PLATO operators' room moved to new locations in the CERL building. Denise cited several reasons for the move, including "better security—both for the offices of the 2nd floor east hallway and for the classroom itself."[108] The most striking reason, however, was that it would be beneficial to have "people (warm bodies) in the immediate vicinity of the ladies' restroom even late at night. (Have had 2 'incidents' that I know of. Thankfully not serious.)"[109]

Denise's euphemism of "incidents" masked the reality that women were physically threatened in a private space designated for them. In another comment on the moves, including changes to restrooms, one PLATO person explained, "Someone was attacked last July at night in the first floor john." Although the person did not specify whether the individual attacked was a man or a woman, additional comments in the thread offered clarity. Another citizen tentatively declared, "Hooray for women's lib?" The question mark emphasized the ambiguity of the situation: good (perhaps?) that the women were receiving another restroom in the building, bad (perhaps?) because the reason had been attacks against women. A final post cemented the casual misogyny of the network and its physical home at CERL: "What's this world coming to? A decent sociopath like myself can't even have a little fun flirting with a chich [*sic*] or two in the restroom without someone getting all excited about it. I guess I can go back to elevators and stairwells." Although it is possible that the author of this post was joking about the situation, the mockery undermined and attempted to minimize the actual problem of violence against women.[110]

The on-line discussion of restroom "incidents" was only one example of how notes contributors emphasized the physical spaces and senses of their computing network. The smells, sounds, and locations of the various PLATO labs mattered enormously to the students and authors who regularly worked on the network. The user "glass" entreated, "How about banning smoking from 2038 and 257????? It would be nice to be able to breath some time while editing," and others "agree[d] completely."[111] The air pollution was often exacerbated by noise pollution. Authors complained about inconsiderate gamers throughout the notes, and Peter bemoaned "the noise generated primarily by game-players (shouts of triumph, dismay, etc.)."[112] Andrew expressed his vote for a "'no noise' and 'priority use of terminals to students' policy WITH enforcement thereof (elimination of records of offending person)."[113] At least two of the PLATO people sought peace and better working conditions in the wee hours of the morning, perhaps contributing to the popular image of the eccentric lone hacker working at all hours. Andrew announced that he was "in nearly every night until early in the morning (say 4 pm to 4 am)."[114] Similarly, author Gerald posted a note at 3:57 a.m. in which he also advocated working during those hours. "True, my sleeping hours are a bit strange," he admitted, "but I can get more accomplished a better percent of the time."[115]

Competition for using the PLATO terminals existed not just on campus at the University of Illinois but in the remote locations too. And simply getting to a PLATO terminal could be challenging at times. Cal exclaimed: "After driving through heavy traffic in Chicago to get to PLATO, and fighting the mobs of people here to get a terminal, and pulling my hair out over not being able to view my lesson in student mode because of ECS [Extended Core Storage, or memory] Crunch, I sure as hell could use some lighthearted humor . . . !"[116] Smoking, noise, traffic—all of these issues reminded the PLATO people of the physicality of participating in their online network. But as Cal intimated, the humor on the network kept drawing people back in.

The notes community enjoyed running jokes, entertaining log-in screens, and even an online newspaper, the *PLATO Press*, but even those recreational features were sometimes debated. The great cookie incident of 1974 combined the dimensions of humor, gender, youth, and the physical nature of PLATO work with, of course, cookies. The cookie thread began when the author Brenda promised cookies to the programmers Anthony and Roger in exchange for the completion of a programming task.[117] Brenda offered to fulfill the traditionally feminine role of baking and food provision in exchange for the masculine task of PLATO programming. In doing so, she inadvertently launched a running joke. Others upped Brenda's cookie ante and proposed "air-mail special delivery," as well as a "bake-off." Later the cookies-for-work discussion expanded to include "the lack of Fanta orange soda Cerl has been experiencing for the past few weeks" as well as pledges for a "case of cold bottles." Another author quickly met the suggestion of beer with a bribe of "Alice B. Toklas [cannabis] cookies," at which point Jeffrey jumped in: "NOW LISTEN ROWELL . . . this system programmer [Roger] is only 14 years old . . . he is being corrupted enough by the computer environment . . . now everyone is paying him off with cookies . . . and let's keep it just that! (give me the other stuff stuff stuff stuff)." After Jeffrey enthusiastically claimed the cannabis cookies for himself, the discussion blossomed. References to cookies, soda, and pizza peppered the notes for months.

The cannabis cookies comment reflected the 1970s' campus culture, as did a PLATO log-in screen that celebrated Illinois's status as the number-one university in the nation for streaking. In 1974, the running-naked fad thrived on campus, including one event with streaking skydivers (naked parachutists), and another with hundreds of naked male student runners and thousands of spectators.[118] A streaker also graced the PLATO log-in screen. In March 1974, when the public clothes-shedding peaked on campus, a PLATO person reported a streaker in the lesson "hypertext," prompting additional discussion and culminating with "a little dude running on the WTP [Welcome to PLATO] page." An author in the military personnel project requested the code for the streaker to put into his course, but another author found him "so frustrating."[119] One scholar has argued that streaking represented a white male "reterritorialization" of campus; that is, the streakers sought to reclaim the campus from the women and minorities who had been empowered during the 1960s.[120]

In light of the sexism of the PLATO network, the computerized streaker may not have been "just" a harmless prank, but rather a sexualized and politicized expression of power over the network. Indeed, this interpretation is bolstered by the fact that an armed services author—someone associated with a preeminent patriarchal organization—wanted the code for his course.

The PLATO people received some special graphics screens with gratitude and little comment, but screens featuring Mickey Mouse and a mail truck sparked extended debate. PLATO featured special displays for Thanksgiving, Chanukah, and Christmas, including a menorah with candles lit on successive days, and a partridge in a pear tree. One comment about the menorah reflected the general jocularity with which such seasonal displays were received: "A true Chanukah miracle is taking place on the 'Welcome to PLATO' page! The menorah is now equipped with _ten_ lights, one burning miraculously in mid-air without a candle to support it! This is especially a miracle, since Chanukah is over!"[121] On the other hand, the appearance of Mickey Mouse on the sign-on screen kindled a heated and extended debate of forty comments (at a time when most notes received one or two responses, a handful at most). People weighed in about whether they liked or disliked the screen, of course, but the appearance of the mouse touched a nerve about the purpose of computing in education. Was PLATO making the students into mindless "IBM"-like "robots in the name of efficiency" above all else, or was PLATO a force for creativity and enjoyment in learning? Author Gerald represented the latter viewpoint:

> What is this, a big monster that grinds in knowledge, or a way to _learn_ (no, Jemimah, not make grades) things? I think that displays like good ol Mickey are the ultimate embodiment of this system—to make doing something something other than a crashing bore! I think perhaps that any anger expressed by students is a reflection of a communications problem . . . a very serious one! Plato is not a surrogate of anything . . . it's a whole nuther thing . . . that's what I try to remember when I'm writing a program.[122]

Gerald believed that PLATO could inspire curiosity, playfulness, and even joy in learning. He also pointed to student frustration or misunderstanding about their use of PLATO, which was a theme articulated by another PLATO citizen:

> There are two issues here (at least) that people are confusing: (1) The mouse is fun, aesthetically pleasing and humanizing. (2) There is much data to support the claim that signing-in is one of the most difficult (confusing) procedures for students. This statement is corroborated by almost every teacher who has run large numbers of students—both on the elementary and university levels. The implication of the above (from my viewpoint and I believe the spirit of Professor [X]'s comments) is that having things like (1) without the problems of (2) is desirable.[123]

Mickey embodied many authors' hopes that PLATO would personalize and even humanize learning, and make it fun. At the same time, because the mouse appeared at sign-on—a process that should have been quick and easy for students but had somehow become arduous—he encapsulated many authors' frustrations with this technology.

The mail truck evoked a similar emotional mix: delight at personalization, dissatisfaction with the time and memory constraints of using a popular, growing, and changing network. Beginning in mid-November 1974, a mailman driving his truck to a mailbox greeted authors at sign-on when they had received a new personal note. One author cheekily suggested that "the emissions device on the mail truck needs fixing—the environment you know," while another called for "an electric-powered one."[124] A week later, the mailman and his truck had disappeared. Some authors complained that he "was not worth the waiting time," but most missed him and vocally expressed their wish for his return. One declared, "SAVE THE MAILNAN!!!!!!!!! [*sic*]."[125] A PLATO person suggested a compromise: "I liked the mail truck and even sent a note to myself once so I could show it to one of my students (have mercy, I was overcome with zeal) but I think the 'surprise' appearance of the truck occasionally (or some other equally entrancing device) would be appropriate."[126] This

PLATO citizen conveyed his amazement and excitement at the novelties of receiving a personal PLATO note and seeing a little mail truck driving across the screen.

The community greeted its own online newspaper, the *PLATO Press*, with less brouhaha than it welcomed Mickey or the mailman. Danny proposed such a newspaper early in 1973. He pitched to his colleagues, "Such a monthly or bimonthly project could be used to provide an information link between Plato users. Although primarily designed on a magazine format, the ability to pool readers would provide a source of information which could be of use to Plato systems people, software development projects, and potential users."[127] Within a year, a systems operator had answered Danny's call and launched the *PLATO Press*. Articles addressed current events, ongoing PLATO research projects, system reliability, and job opportunities. When the much-anticipated comet Kohoutek appeared over Earth for the first time in 150,000 years, the *Press* "made it easy to find" for readers.[128] When Champaign-Urbana celebrated its own "World Plan Week" to promote transcendental meditation (TM) and relieve world suffering, the *Press* offered "scientific studies and tests on the physiological, psychological and sociological benefits of TM."[129] One PLATO person objected to "all that nonsense" about TM, but others contended that "articles of this type should not be excluded just because they do not directly bear upon plato."[130] Most seemed to favor any articles of potential interest to the large and far-flung PLATO community. The *Press* featured interviews with key PLATO people, including "the man" Don Bitzer and the programmer Andrew, who created a game of checkers for the system.[131] Andrew later mused, "I don't think the press did me in anywhere near as badly as one normally expects!"[132] After the first few issues, the programmer who had launched the *Press* posted enthusiastically, "This seems to be a record breaking edition!" with several hundred readers in two days.[133]

The debates over the contents of the *PLATO Press* formed part of a larger conversation about censorship on the PLATO network. The PLATO group was generally flexible on the topics allowed on their system, but politics became a gray area. Stuart Umpleby and others had posted statements about the impeachment of President Richard Nixon on the ARPANET, and their actions were covered by the journal *Change*. In a letter to the editor clarifying the situation, and contrasting the ARPA network with PLATO, Umpleby explained that on PLATO, "anything that might be discussed in a classroom can be discussed on the system, but political organizing is not allowed."[134] Umpleby went a step further and suggested that PLATO's guidelines also be applied to discussion on the ARPANET, rather than allowing the Department of Defense to have the final say on ARPA network censorship.

Denenberg also commented on censorship (or lack thereof) on the PLATO system. He noticed that "the default position in the university environment has been to rely whenever possible on peer group pressure to regulate the quality and the content of notes and lessons; in cases where people are acquainted with one another, this pressure seems to work adequately well."[135] Denenberg thought that "use of language" posed little to no problem on the PLATO system, and he observed that a notesfile called "sexforum" existed "more or less benignly" on PLATO.[136] Here, Denenberg's gender may have skewed his perspective. One wonders if the "use of language" troubled any minority groups on the PLATO system (whether women, or sexual or racial minorities), and how others regarded the "sexforum" file. Indeed, at one point, the language on PLATO bothered Frank enough for him to post: "To systems people: the game moonwar has many possibilities to abuse the message function as many people are using it for message[s] that I would not want my children seeing."[137]

Games like *Moonwar*, in which Frank had found obscene messages, were indubitably the most highly contested of the PLATO realms. PLATO games on the system by 1977 included checkers, chess, blackjack, solitaire, maze games, *Physgame* (physics games), *West* (a numbers game), *Game of Life*, *Moonbattle*, *Moonwar*, *Speedway*, and *Empire*. Some of those games were played against the computer, but others, like *Moonwar* or *Empire*, were two-player or multiplayer games that took advantage of PLATO's network. In those two-player or multiplayer games, users who were physically miles away from each other could participate in the same game at the same time. Hierarchies clearly emerged, and users even took matters into their own hands.

Simply stated, many PLATO people (from students to system programmers) liked playing games. Others, typically authors, consultants, and programmers, resented the gamers, especially when they were trying to program PLATO lessons. Denenberg thought that "the management attitude towards games was essentially negative. Games were viewed as frivolous, a waste of time (no 'learning' was occurring), and a waste of scarce computer resources. As a result, although games were often used in demonstrations, they were in fact heavily discouraged at all levels of management."[138]

Games remained on the network for demonstration purposes, and people loved them, but there was constant debate about who could play the games, and when. Perhaps most importantly, authors wondered whether gamers should vacate their terminals for nongaming users. In January 1973, Carol posted a "GREAT GRIPE!" that for two nights in a row, in the main classroom, people were playing games on all eighty terminals.[139] The next day, another PLATO person posted a long and heartfelt response to Carol's complaint, in defense of games. A few weeks later, system administrator Anthony reported that he had "sabotaged as many games" as he could find "after the morons using them refused to stop after repeated requests via message."[140] Most of the gamers were students, and this created its own confusion because there was internal disagreement among the PLATO team (consultants, programmers, systems administrators) about the priority of students on the system compared with authors. Denenberg noticed the "management attitude of 'students can't really be trusted' and a sense of a definite power hierarchy of student/author/course director/PLATO project programmer (each had more secret codes than the next whereby he could unlock more of the information within the system)."[141]

The debate over games continued for at least another year, and when it emerged again in February 1974, PLATO staff member Kenneth articulated a very "hands-off" policy: "The PLATO staff are fully aware of the educational, recreational, and other values of games. We have never considered taking game lessons off PLATO. Having said that, I wish to add that we cannot condone game playing in rooms 2038 or 257 which annoys students or authors!"[142] Kenneth also admitted that the staff could not really enforce their rules, and the debates around games continued. Some authors considered their work far more useful than the games, while the game advocates continued to defend their actions. At one point, a PLATO citizen even suggested, "Those PLATO users that consider 'game' lessons to be of less value than other uses could perhaps benefit from reading M_an and the Computer_ [*sic*] by J. G. Kemeny, President of Dartmouth College, 'father of timesharing,' and co-inventor of BASIC."[143]

Conclusion

In 1973, Bitzer had penned an article in which he described the future. He wrote about a young boy touching a computer screen to hear the pronunciation of a word that he could not read, or to see a picture of that word. Bitzer portrayed a woman playing a game of bridge with the computer, while the computer coached her on how to improve her bidding techniques. He depicted a businessman about to jet off to Paris for an important meeting, practicing key French phrases with the computer, complete with pronunciation review.[144] Bitzer knew that his scenarios seemed like the distant future to almost everyone, and he enjoyed publicizing that his PLATO network could already perform all of these individualized, interactive—personal—tasks. For Bitzer, the plasma screen had been the heart of his PLATO IV system, and the plasma screen had propelled the development of a network with hundreds of terminals across the United States. Yet, once it was in their hands, the users made PLATO IV their own. They embraced PLATO as a communications medium and as a recreation platform. They wrote lessons for their students, but they also shared jokes and news, and received advice on bicycle maintenance. They procrastinated by reading and posting to the notes files, they met online, they courted and flirted, and they got married.[145] PLATO users lived and breathed a digital culture—one that looked very much like the future would.

Epilogue

From Personal Computing to Personal Computers

I HAVE ARGUED that 1965–1975 was a golden age of networked computing, a decade during which students, educators, and enthusiasts created personal and social computing before personal computers. Their computing access went hand in hand with network access. They formed their networks into commons in which they shared programs, helpful hints, cautions, ideas, and experiments. I contend that they were computing citizens, rather than computing consumers. They were computing citizens because their computing access was subsidized by the communal institutions of which they were members, whether universities, colleges, or K–12 schools. They were citizens because they forged new computing communities, and new social and technological networks with their access and participation. They were not consumers because they did not have to purchase their computers. They were also not consumers in the sense of end users. They did not passively use programs and software that had already been produced by someone else. They learned BASIC (or TUTOR). They wrote new programs for novel and personal purposes. They created poetry, music, activism, and art.

The federal government, via the National Science Foundation (NSF), and the educational institutions (themselves supported by local, state, and national funding) underwrote this golden age of networking, an investment that too often has been overlooked; nonetheless, businesses did contribute in varying degrees to the Kiewit, BASIC (Beginners' All-purpose Symbolic Instruction Code), Minnesota, and PLATO (Programmed Logic for Automatic Teaching Operations) networks. GE profited from Dartmouth's time-sharing expertise, while minicomputer manufacturers like Digital Equipment Corporation (DEC) and Hewlett-Packard (HP) monetized their support of educational materials, like the Huntington Project, to sell their machines. Control Data Corporation (CDC), headquartered in Minnesota, provided a computer to helm Minnesota's statewide network and had a hand in PLATO's success.

In the contemporary popular American narrative of computing and networking, players such as GE, DEC, and CDC have been forgotten, not to mention people such as Kemeny and Kurtz; Albrecht and Braun; LaFrenz, Borry, and Danver; and Bitzer, Lamont, and Umpleby. That popular narrative jumps from the oppression of 1960s mainframes to the liberation of personal computers in 1975–1985 to the glorious unification and diversification of the 1990s Internet (perhaps now with a turn to the freedom of 2000s smartphones).

Instead, I have shown that personal and social networked computing thrived before personal computers. In my narrative, the post-1975 turn to personal computers represents a time in which possibilities were foreclosed, connections were severed, and computing communities waned. Indeed, following the threads of PLATO, BASIC, and MECC (Minnesota Educational Computing Consortium) from 1975 into the 1980s reveals the great losses in the shift from personal computing to personal computers.

PLATO: Control Data's Struggle to Sell

PLATO and CDC mutually benefited from a long-standing relationship. In the early 1960s, while PLATO was still running on the ILLIAC machine at the University of Illinois Coordinated Science Laboratory (CSL), the fledgling PLATO team perpetually elbowed their way into more time on the lab's powerful new CDC machine. Bitzer recalled that the CSL purchased the CDC machine because the company had a particularly effective salesperson.[1] Bitzer's recollection may have reflected the bias of his, by 1988, nearly three-decade relationship with CDC; regardless, he and the early PLATO people received exposure to CDC computing early on. The CSL obtained a CDC-1604 computer during the summer of 1961, and competition for computing time intensified.[2] Although both military and university representatives had found demonstrations of PLATO I quite compelling and urged Bitzer to proceed, Bitzer recalled that other CSL staff viewed PLATO as frivolous. He related, "The fourth floor [CSL] people who were involved in air defense, sea and air defense, thought that we were eating up too much of their resources. And they were responsible for the computer. That was the big problem."[3] Indeed, Bitzer's need to prove himself to his peers at CSL—and later to the university, American public education, and the computing industry—may well have propelled his ambition for and his devotion to PLATO.

The installation of the CDC-1604 meant rewriting PLATO II for a new computing system, but it also signaled the beginning of a PLATO-CDC relationship that would span three decades. Bitzer recalled that immediately after the CDC-1604 began running at CSL, the most frequently run program was SIMILLIAC—a simulator for the ILLIAC.[4] Running PLATO II on the ILLIAC and on the CDC-1604 in SIMILLIAC mode enabled Bitzer and his colleagues to increase their PLATO computing time. The dynamic PLATO team attracted the attention of Harold Brooks, the sales representative from CDC. Brooks observed the PLATO project with interest, visiting Bitzer and his team when he was in town. Bitzer explained that, over time, Brooks learned about the PLATO team's challenge of obtaining more computing time.[5] Ultimately, Bitzer traveled to CDC in Minnesota to meet with the company's cofounder and president, William (Bill) C. Norris.[6] They arranged for the PLATO project to lease a CDC-1604 essentially for free; the project had to cover only the cost of computer maintenance and upkeep.[7] This was a sound investment for CDC. Norris recognized a potentially large and lucrative market in computer-based education, and he was aware that competitors like IBM were investing in that market.[8] Norris believed in the promise of PLATO, and his gift of a CDC computer signaled the beginning of his own long-term personal relationship with the system.

Bitzer and Norris further deepened the PLATO-CDC relationship later in the decade when CDC sold another computer to PLATO's home laboratory (Computer-based Education Research Laboratory [CERL]), followed by additional contracts during the 1970s and 1980s; meanwhile, Norris started devoting resources to commercializing PLATO for CDC. In 1971 CDC began developing its own PLATO courses, including employee training materials, within its new Education Department. Three years later, CDC installed its own PLATO IV system at its Minnesota headquarters. In 1976, the University of Illinois licensed to CDC the full rights to sell all of PLATO—hardware including the plasma display screens, software including the networked communications options, and courses developed at Illinois. CERL (PLATO's home) retained the rights to market PLATO to customers who desired a maximum of four terminals. After the agreement, Bitzer and CDC continued their relationship. Bitzer demonstrated PLATO at CERL to potential CDC clients, and he even traveled to Africa, Australia, Europe, and South America to promote PLATO for CDC.[9]

Two PLATO worlds emerged: the system centered at CERL and the one controlled by CDC. During the 1970s, one observer with experience at both contrasted these diverging systems. Echoing user comments from the notes that we saw in Chapter 7, he lamented that CERL discouraged PLATO game playing. He celebrated CDC for promoting a PLATO gaming culture, although he also acknowledged that CDC ultimately sought to sell PLATO, and games were an engaging and compelling demonstration of the system's capabilities. He explained, "After 5 P.M., however, I noticed that several programmers stayed on to relax and unwind by playing some of the more aggressive games such as 'airfight.' Games seem to be viewed as healthy and useful at CDC while they are considered, at best, a necessary evil in academia."[10] The author, who had almost a decade of experience with PLATO as a user, instructor, programmer, and consultant, also reflected that corporate CDC PLATO was far more concerned with potentially offensive lessons or notes files. He categorized this as a problem of censorship; however, such caution could also have been attributed to the fact that CDC was attempting to sell education—at all levels from elementary to adult. Thus, CDC had to be considerate of the language, comments, and topics that might chase away potential buyers. The parallel PLATOs ultimately shared little courseware; CDC worked to develop many of its own courses.[11] Here again, the "personal" evaluator of both systems offered some insight. He claimed that any PLATO course was only as good as its developer, and he decried the bad ones as "worse than useless."[12] He explained that the PLATO course programming language, TUTOR, eased the creation of simple programs (like drill and practice), but TUTOR made it "extremely difficult for anyone to write (and especially debug) a moderately complex (and useful) program."[13] Ultimately, he concluded that only 5–10 percent of all PLATO lessons (at both CDC and the University of Illinois) were valuable and instructive.

CDC president Norris poured millions of dollars into PLATO, and he confidently declared that its revenues would make up half of CDC's business by 1985; yet CDC PLATO struggled. As soon as the licensing agreements were in place with the University of Illinois, CDC began marketing several PLATO purchase options: buy a complete PLATO system with CDC mainframe, terminals, and courses; buy some terminals connected to CDC's own mainframe; or buy a single PLATO course at one of the company's various learning institutes around the world. CDC was trying to sell education, but studies—including one by Educational Testing Services in 1977—began showing that PLATO performed no better than a human teacher. CDC was trying to sell courses, but even by 1975, many of PLATO's tacitly recognized benefits were those of the network: communication files, screen sharing, instant messaging, other forms of terminal-to-terminal communication, and working in a social group setting (even while pursuing individual work). By 1980, the year in which CDC introduced Micro-PLATO, the company had spent $600 million on PLATO. Micro-PLATO, CDC's attempt to jump on the microcomputer bandwagon, flopped. Micro-PLATO severed network connections and offered PLATO courses on floppy disk, thereby eliminating network benefits, such as communication or online support.

Using a CDC-branded PLATO touch-screen terminal. Note the "Control Data" logo on the lower left corner of the screen. Image courtesy of the Charles Babbage Institute, University of Minnesota Libraries, Minneapolis, Minnesota.

By March 1982, CDC had dumped $900 million into PLATO with no substantial profit. Norris retired in 1986, and three years later his successor sold the entire PLATO division.[14]

PLATO thrived at Illinois; it sank CDC—what happened? CDC was selling PLATO. Its goal was to create PLATO consumers. In contrast, at the University of Illinois, the PLATO people were citizens; their participation on the system was subsidized by institutional funding from the Advanced Research Projects Agency (ARPA), the NSF, the university itself, and more. CDC PLATO focused on selling to the student as the end user; the product was essentially computer courses. In contrast, although the Illinois PLATO system was purportedly dedicated to the student, its most active—and privileged—citizens were the CERL staff, PLATO programmers and consultants, and course authors who regularly availed themselves of PLATO's network features: notes files, screen sharing, instant messaging, and

Students eagerly gather around a Control Data PLATO terminal in this photograph used in a 1980 advertising campaign. Image courtesy of the Charles Babbage Institute, University of Minnesota Libraries, Minneapolis, Minnesota.

multiplayer gaming. The Illinois PLATO people loved their computing-based community; their personal and social computing had been an unexpected boon from the education-oriented system. CDC PLATO may have internally resembled that rich community culture; however, what CDC was selling in PLATO was not personal or social computing; it was just courses.

From BASIC Personal Computing to BASIC Personal Computers

Recall from Chapter 3 that in 1972 Bob Albrecht had established the *People's Computer Company*, a periodical dedicated to providing people with information about personalizing their computing experiences. The first issue proclaimed, "The *People's Computer Company* is a newspaper . . . about having fun with computers, and learning how to use computers, and how to buy a minicomputer for yourself or your school, and books . . . and films . . . and tools of the future."[15] With its mesmerizing mix of fonts, teletype printouts, cartoons, illustrations, and photographs, the *People's Computer Company* reached a circulation of eight thousand.[16] The *People's Computer Company* proselytized BASIC. The newspaper published games in BASIC so readers could create the games on their own computing systems. It regularly offered information on the DEC and HP minicomputers that "spoke" BASIC, as well as books and other resources for learning BASIC. Albrecht also founded the associated People's Computer Center, a storefront that offered public computing access. Albrecht and the *People's Computer Company* were central to the Bay Area home computer endeavor; in fact, one of Albrecht's *People's Computer Company* instructors, Fred Moore, founded the Homebrew Computer Club.[17]

When MITS (Micro Instrumentation and Telemetry Systems) publicized its Altair 8800 microcomputer in 1975, Albrecht and the *People's Computer Company* recognized its "home computer" promise; a *People's Computer Company* cover featured the Altair, but more importantly, the *People's Computer Company* recognized the need for a BASIC for the Altair.[18] In the issue immediately following its Altair-cover issue, the *People's Computer Company* put out a call to its community of computing people. Under the heading "Build Your Own BASIC," they proposed a pared-down language called "TINY BASIC," and urged, "We'd like everyone interested to participate in the design. . . . Your thoughts, ideas, etc. about TINY BASIC urgently requested."[19] Albrecht and other *People's Computer Company* contributors recognized that the "home computer" had "not too much memory," and they proposed Tiny BASIC to overcome that limitation.[20]

The *People's Computer Company*'s request was well met; a communal effort produced Tiny BASIC and inspired a new publication, the *Dr. Dobb's Journal of Computer Calisthenics and Orthodontia . . . Running Light without Overbyte*. *People's Computer Company* readers enthusiastically tinkered with Tiny BASIC, refined it, and reported new dialects, games, and other interesting programs. They freely shared their experience and expertise. Albrecht decided Tiny BASIC merited its own publication. The quirky name came from a combination of Dennis (Allison) and Bob (Albrecht), who had launched Tiny BASIC, with the understanding that as "an exercise in computer programming," BASIC was "calisthenics" and because Tiny BASIC did not "use very many bytes of memory," it was "avoiding overbite." All together: *Dr. Dobb's Journal of Computer Calisthenics and Orthodontia* (published until 2014 as *Dr. Dobb's Journal*).[21] In its first issue, the editor of *Dr. Dobb's Journal* clearly stated his intent for a "communication medium concerning the design, development, and distribution of free and low-cost software for the home computer."[22] In other words, just as *People's Computer Company* had promulgated computing of the people and for the people, so too would *Dr. Dobb's Journal.* And just as BASIC was vital to the era of personal computing before personal computers, Tiny BASIC became vital to the era of hobbyists and home computers.

People's Computer Company and *Dr. Dobb's Journal* espoused communal computing built on BASIC and Tiny BASIC, respectively; *Dr. Dobb's* in particular presented itself as a "viable alternative to the problems raised by Bill Gates in his irate letter to computer hobbyists concerning 'ripping off' software."[23] And so BASIC brings us to Bill Gates. Gates gained formative computing experience in middle school working on a teletype connected to a GE time-sharing computer.[24] Recall from Chapter 4 that GE had built its computing business—both the manufacturing of time-sharing computers and selling time-sharing as a utility—based on its collaboration with Dartmouth College. Because Dartmouth implemented BASIC with such success, GE followed suit. And so a young Bill Gates owed his initial computing experience to that particular combination of Dartmouth, GE, time-sharing, and BASIC.

While Albrecht and the *People's Computer Company* had crowdsourced a BASIC for the Altair home computer, Gates and his high school friend Paul Allen pitched the sale of a BASIC interpreter directly to the Altair manufacturer, MITS.[25] Shortly thereafter, they founded Microsoft (a combination of microprocessors and software) to sell software to different computer vendors. Gates recalled,

> I was doing the payroll, writing the taxes, doing the contracts, figuring out how to price the software. In fact, I was business-oriented enough that I wrote a letter about software piracy, sort of complaining that a lot of these computer groups weren't paying for their software. That really became a *cause celebre* at

> the time: "Is it fair that this guy is asking for money? Should we pay for this stuff?"[26]

In his February 3, 1976 letter, Gates bemoaned the "lack of good software courses, books and software itself" in "the hobby market."[27] He then indicted many of his fellow computing enthusiasts: "As the majority of hobbyists must be aware, most of you steal your software. Hardware must be paid for, but software is something to share."

Gates's indictment of the people computing contained a problematic conflation—one that ignored BASIC's previous decade; he equated sharing software with stealing software. Yet Kemeny and Kurtz had freely shared BASIC on purpose. They wanted to spread a language that encouraged personal computing. From 1965 through 1975, BASIC programs, including those of the Huntington Project and the *People's Computer Company*, circulated widely at little or no cost. Gates did articulate the time-consuming nature of writing and perfecting such programs; however, many others—like the editor of *Dr. Dobb's Journal*—believed that "when software is free, or so inexpensive that it's easier to pay for it than to duplicate it, then it won't be 'stolen.'"[28] Although the editor of *Dr. Dobb's Journal* differentiated between "business and industrial communities" and the "hobby" or "academic environment," and esteemed the sharing ethic of the latter two, Bill Gates elevated selling over sharing. Gates's Microsoft initially sold software to other businesses, but it eventually sold software directly to consumers. Gates's letter marked the start of a shift from people computing to people consuming. Increasingly, people would have to purchase computers and software (now, devices and apps) for their personal and social computing.

BASIC also figures prominently in the history of Apple. Steve Wozniak produced his own "Integer BASIC" for his homemade computer, built around MOS Technology's 6502 microprocessor chip; he shared Integer BASIC, and he even published programs in *Dr. Dobb's Journal*.[29] When Wozniak's high school chum Steve Jobs saw the computer, he proposed they team up to assemble and sell them. They named the computer Apple, and soon began working on a new version, the Apple II. Although Apple declared its philosophy was "to provide software for our machines free or at minimal cost," Apple sought (aggressively) to sell its hardware.[30]

Whether they were called home computers, hobby computers, microcomputers, or personal computers, they were consumer products, purveyed by Steve Jobs. Ted Nelson recognized this in his keynote address during the first West Coast Computer Faire. Although Nelson was excited about the "magic" of small computers, he also recognized the dollar signs driving the fair's frenetic energy. He blazed, "The little computers are here, you can buy them on your plastic charge card, and the available accessories include disc storage, graphical displays, interactive games . . . and goodness knows what else. . . . FAD! CULT! CONSUMER MARKET! The rush will be on. The American manufacturing publicity machine will go ape. American society will go out of its gourd."[31]

MECC: From Social Networks to Isolated Computers

While researching this book, even before I started writing, I realized I had an ending in mind. The cover of MECC's November–December 1978 *Dataline* newsletter trumpeted, "State microcomputer bid awarded to Apple Computer, Inc." The article featured a photo of Linda Borry, who had computed music programs and worked for TIES (Total Information for Educational Systems) and then MECC, using the "Apple II microcomputer."[32]

To me, MECC's 1978 contract with Apple symbolized the transition from every student having a network to every student having a computer. The student gained access to a personal computer, but she lost access to her computing network. At the same time, MECC gradually shifted from offering programs as part of the commons available on its network (and subsidized by the state as such) to selling its software.

In the year leading up to its partnership with Apple, MECC proudly reported on the reach of its time-sharing network. Don Rawitsch, one of the designers of *The Oregon Trail*, "emphasize[d] user services" for the community college system.[33] Altogether, the statewide network reached 310 school districts, 19 community colleges, and 7 statewide universities, with 1,500 computing terminals serving an estimated

The photograph of Linda Borry with an Apple II that appeared on the cover of MECC's *Dataline* newsletter for November–December 1978. Courtesy of the Minnesota History Center, Minnesota Historical Society.

400,000 students. MECC people reflected on their "awareness and appreciation" of computing, just as they praised the social aspects of their network, the rich "people-to-people contact."[34] Networked computing via time-sharing thrived in Minnesota, having been motivated by a desire to equally serve citizens across the state, to provide the "same opportunities in computing . . . to a rural resident as to anyone in the suburbs or Twin Cities."[35]

Just as DEC had spurred sales of its time-sharing minicomputer systems by making them attractive to students and educators, Apple also targeted schools in the late 1970s. Steve Jobs later explained that "schools buying Apple IIs" was "one of the things that built Apple IIs."[36] Indeed, a MECC staff member who had attended a conference in California reported to his colleagues in Minnesota about the amazing Apple II computer he had seen, and MECC soon arranged to buy over five hundred of them from Jobs and Wozniak.[37]

MECC led the nation in placing microcomputers in its classrooms, and Apple gained an early and large share of the educational computing market; moreover, the long-ago decision to implement Dartmouth Time-Sharing with BASIC at UHigh in Minneapolis had important ramifications all these years later.[38] MECC and its constituent members (such as TIES and MERITSS [Minnesota Educational

Regional Interactive Time-Sharing System]) had been building a library of applications and games—written in BASIC—since the 1960s.

When MECC adopted Apple IIs, the consortium gained a new role, that of software translator from time-sharing BASIC to Apple BASIC.[39] When Apple began selling a rapidly increasing number of microcomputers to educational systems around the country, teachers in those schools soon realized they could not accomplish much with the machines alone. They needed applications, and MECC gained a reputation for its software library; its diskette collection was "in demand from educators around the world."[40] The consortium soon developed a distribution system by which other organizations could purchase its software and share it with a specified group of users—a novel "membership" mode of operation in the computing business.[41] By 1982 MECC's members included schools across the United States and in Australia, England, Kenya, Saudi Arabia, and Switzerland.[42] MECC's early adoption of Apple hardware, paired with MECC's extensive software offerings, gave Apple a strong foundation in selling to students, their parents, and educators in the early 1980s.

MECC continued to operate its time-sharing network into the 1980s, in no small part because the Apples were not very reliable. In 1981 MECC's director of instructional services argued that time-sharing was "the backbone of the instructional computing. . . . I don't believe we should throw out the time-share system." He acknowledged the "mental set that says—'Microcomputers are it'"; however, he lamented, "Right now, more than one-half of the Apples that MECC has purchased are in for repair." That meant hundreds, if not thousands, of the MECC's educational computers were not serving students and educators, not affording them computing opportunities. The director bemoaned this "significant impact" on Minnesotans, alongside the high cost of maintaining the Apples.[43]

Although MECC had been established to cultivate computing citizenship equally across Minnesota, MECC ultimately turned the resources of its extensive interpersonal and computing networks toward consumer ends; in 1983 the state rechartered MECC as a for-profit taxable organization. For the rest of the decade, MECC's revenues climbed into the millions of dollars, propelled by the wild success of *The Oregon Trail.* During the mid-1990s, with annual revenues approaching $30 million, Minnesota sold MECC to a venture-capital firm. MECC, once a model provider of personal computing before personal computers, had become an emblem of the consumer market.

The people computing became the people consuming. Rather than gaining computing through schools—or through an envisioned public computing utility—Americans had to buy a product, a personal computer. Rather than paying for a malleable computing utility capable of connecting people through civic, social networks, Americans had to pay for isolated, individual machines and the floppy disks that animated them. The ability to own, possess, and consume a device became paramount.[44]

There was a shift from a focus on users, the network, and the communal to a focus on the user, the machine, and the personal. Analyzing the creativity, collaboration, and community of 1960s and 1970s academic time-sharing networks enables the distinction between the computing citizens of the 1960s and 1970s and the computing consumers of the 1980s and beyond.

I have argued that we saw networked social computing as early as the 1960s. From time-sharing computing networks to various regional computing networks, then to the Bulletin Board Systems (BBSs) of the 1980s, including the well-known WELL (Whole Earth 'Lectronic Link) and the lesser-known ECHO (East Coast Hang Out), there have been connections and continuities.[45] If we think about the continuities of computing networks from the 1960s onward, and about the ages and experiences of the individuals using those networks, we can create a much richer portrait of our digital nation.

From the perspective of continuity, the experiences that *New York Times* critic Virginia Heffernan recounts in her book *Magic and Loss: The Internet as Art* no longer seem an exotic outlier.[46] Heffernan describes "stumble[ing] onto the early Internet" around 1980 as a ten- or eleven-year-old girl—by connecting to the Dartmouth Time-Sharing System.[47] She loved the live chatroom that she identifies as Conference XYZ. Heffernan observes, "The story of early computer networks has most often been told as a technology and business story. But like the Internet today, Conference XYZ was not an engineering experiment as much as an immersive experience. . . . While we were seeking connection and community we were also helping to build a culture. Today I see that culture writ large online."[48] I, too, see that "culture writ large online," and *A People's History of Computing in the United States* locates the origins of that online culture in the 1960s with other ten- and eleven-year-old girls and boys, other high school teachers, and other college students.

I have written *A People's History of Computing in the United States* with the hopes that it will be the first of many such histories. This is emphatically "a" history, one possible out of many.[49] I have written this book not to be exhaustive, but to be definitive. There was an amazing world of personal computing, social computing, and networked computing—all before 1975—and there is so much more to learn about how those worlds became the American digital culture that we recognize today. My most sincere hope is that this book inspires tens of others. Digital culture is no longer inseparable from American culture; the history of that culture should no longer be treated as separate from American history. Let us overwrite the Silicon Valley mythology. Let us look beyond the narrow inspiration of digital Founding Fathers. Let us recover the many people's histories of computing.

Notes

Bibliography

Acknowledgments

Index

Notes

Introduction

1. G. Albert Higgins, "BASIC: A New Language at Mount Hermon," *The Bulletin of Northfield and Mount Hermon Schools* (Spring 1967), clipping in Folder DA 29 (7841) 5 of 6, labeled: Miscellaneous Time-Sharing, Dartmouth College Rauner Special Collections Library (hereafter Rauner Library).

2. Gordon D. Goldstein, Laura A. Repass, and Barbara J. Walker, *Digital Computer Newsletter* 20, no. 1 (Arlington, VA: Office of Naval Research, January 1968), http://www.dtic.mil/docs/citations/AD0694655.

3. During the 1980s, historians including William Aspray, James Cortada, Paul Ceruzzi, and Michael Mahoney established an early and influential vision for the history of computing that focused on men, machines, mathematical theories, and businesses. Refer to the Bibliography for selected examples of their scholarship. A more recent popular account is Walter Isaacson, *The Innovators: How a Group of Hackers, Geniuses, and Geeks Created the Digital Revolution* (New York: Simon & Schuster, 2015).

4. The most prominent account positioning computing as a "Cold War technology" is Paul Edwards's *The Closed World: Computers and the Politics of Discourse in Cold War America* (Cambridge, MA: MIT Press, 1996). Others suggest that the generous funding of the Cold War research economy allowed for creativity and a vibrant exchange of ideas. See, especially, Arthur Norberg, Judy O'Neill, and Kerry Freedman, *Transforming Computer Technology: Information Processing for the Pentagon, 1962–1986* (Baltimore: Johns Hopkins University Press, 1996) and Atsushi Akera, *Calculating a Natural World: Scientists, Engineers, and Computers during the Rise of U.S. Cold War Research* (Cambridge, MA: MIT Press, 2007). Accounts that highlight the role of the counterculture in personal computing's origins include Fred Turner's *From Counterculture to Cyberculture: Stewart Brand, the Whole Earth Network, and the Rise of Digital Utopianism* (Chicago: University of Chicago Press, 2006) and John Markoff's *What the Dormouse Said: How the Sixties Counterculture Shaped the Personal Computer Industry* (New York: Viking Penguin, 2005).

5. These works include Susan Rosegrant and David Lampe, *Route 128: Lessons from Boston's High-Tech Community* (New York: Basic Books, 1992);

AnnaLee Saxenian, *Regional Advantage: Culture and Competition in Silicon Valley and Route 128* (Cambridge, MA: Harvard University Press, 1994); Christophe Lécuyer, *Making Silicon Valley: Innovation and the Growth of High Tech, 1930–1970* (Cambridge, MA: MIT Press, 2006); and Barry Katz, *Make It New: The History of Silicon Valley Design* (Cambridge, MA: MIT Press, 2015). Margaret Pugh O'Mara's *Cities of Knowledge: Cold War Science and the Search for the Next Silicon Valley* (Princeton, NJ: Princeton University Press, 2005) includes Philadelphia and Atlanta, but offers a chapter on Silicon Valley as the basis for comparison. Notable exceptions to this American bicoastal focus are Paul Ceruzzi's *Internet Alley: High Technology in Tysons Corner, 1945–2005* (Cambridge, MA: MIT Press, 2008) and Thomas Misa's *Digital State: The Story of Minnesota's Computing Industry* (Minneapolis: University of Minnesota Press, 2013).

6. Examples include Thomas Mullaney's *The Chinese Typewriter* (Cambridge, MA: MIT Press, 2017); Alison Gazzard's *Now the Chips Are Down: The BBC Micro* (Cambridge, MA: MIT Press, 2016), about Great Britain; Julien Mailland and Kevin Driscoll's *Minitel: Welcome to the Internet* (Cambridge, MA: MIT Press, 2017), about France; *Connecting Women: Women, Gender and ICT in Europe in the Nineteenth and Twentieth Century*, ed. Valérie Schafer and Benjamin Thierry (Cham: Springer, 2015); Dinesh Sharma's *The Outsourcer: The Story of India's IT Revolution* (Cambridge, MA: MIT Press, 2015), about India; and Benjamin Peters's *How Not to Network a Nation: The Uneasy History of the Soviet Internet* (Cambridge, MA: MIT Press, 2016).

7. See, for example, David Mindell's studies of engineers (*Between Human and Machine: Feedback, Control, and Computing before Cybernetics* [Baltimore: Johns Hopkins University Press, 2002]) and aviators (*Digital Apollo: Human and Machine in Spaceflight* [Cambridge, MA: MIT Press, 2008]). The following works emphasize gender: Nathan Ensmenger, *The Computer Boys Take Over: Computers, Programmers, and the Politics of Technical Expertise* (Cambridge, MA: MIT Press, 2010); Janet Abbate, *Recoding Gender: Women's Changing Participation in Computing* (Cambridge, MA: MIT Press, 2012); and Marie Hicks, *Programmed Inequality: How Britain Discarded Women Technologists and Lost Its Edge in Computing* (Cambridge, MA: MIT Press, 2017).

8. Some readers may argue that time-sharing systems were not "true" networks. While they were certainly not networks in the sense of the contemporary Internet, time-sharing systems most certainly were networks, and they were recognized as such at the time, as abundant examples throughout the rest of the book demonstrate. One such example is a circa 1970 brochure, *The Kiewit Computation Center & The Dartmouth Time-Sharing System*, featuring a map of the "Dartmouth Educational Time-Sharing Network" in the Information File for Time-Sharing, Rauner Library.

9. Multiplayer games in *Kiewit Comments* 3, no. 7 (November 4, 1969), MAILBOX in *Kiewit Comments* 2, no. 6 (August 19, 1968), Rauner Library.

10. I must underscore the methodological value of studying online systems that originated in an educational context. Because many of the project publications were written for readers in education, they often included meticulous details of individuals' encounters with the terminal, the keyboard, the screen, the language, the lessons, the appropriate syntax, and similar issues. The re-

ports documented knowledge and practices that otherwise would have been tacit and unnoticed. This is immensely helpful for historians seeking to understand and describe novel experiences of computing and networking.

11. In challenging the boundaries between users and producers, I have been inspired by critical scholarship in the field of science and technology studies. See especially *How Users Matter: The Co-Construction of Users and Technologies*, ed. Nelly Oudshoorn and Trevor Pinch (Cambridge, MA: MIT Press, 2003), as well as Alan Irwin's elaboration of "scientific citizenship" in "Constructing the Scientific Citizen," *Public Understanding of Science* 10, no. 1 (2001): 1–18, and "Citizen Science and Scientific Citizenship," in *Science Communication Today—2015*, ed. B. Schiele, J. L. Marec, and P. Baranger (Nancy, France: Presses Universitaires de Nancy, 2015), 29–38. Advocates of 1960s and 1970s time-sharing networks often shared the belief that computing, information, and knowledge were becoming increasingly crucial to American economic and social success; that is, computing would be essential to the nascent knowledge society.

12. I appreciate Margarita (Margo) Boenig-Liptsin's attention to citizenship in "Making Citizens of the Information Age: A Comparative Study of the First Computer Literacy Programs for Children in the United States, France, and the Soviet Union, 1970–1990" (PhD diss., Harvard University, 2015).

13. As I explain in my essay "Toward a History of Social Computing: Children, Classrooms, Campuses, and Communities," *IEEE Annals of the History of Computing* 36, no. 2 (2014): 88, 86–87, I employ the phrase "social computing" to emphasize the social connections forged around and with computing use. This history of social computing is informed by traditions of social history that are now well established in American history and are becoming more common in the history of science. For example, in *The Jewel House* (New Haven, CT: Yale University Press, 2007), Deborah Harkness locates a Scientific Revolution percolating in the streets of Elizabethan London. See also Nathan Ensmenger, "Power to the People: Toward a Social History of Computing," *IEEE Annals of the History of Computing* 26, no. 1 (2004): 96, 94–95.

14. The collection coedited by the political theorist Danielle Allen and the historian Jennifer S. Light, *From Voice to Influence: Understanding Citizenship in a Digital Age* (Chicago: University of Chicago Press, 2015), demonstrates the value of taking a historical approach to the relationships between citizenship and digital media.

1 When Students Taught the Computer

1. Thomas E. Kurtz, oral history interview by Daniel Daily in Hanover, NH, on June 20, 2002, and July 2, 2002, DOH-44, Dartmouth College Rauner Special Collections Library (hereafter Rauner Library); Thomas E. Kurtz, Leonard M. Rieser, and John F. Meck, "Application to the National Science Foundation Mathematics Division for the Establishment and Support of the Dartmouth Computation Center, 22 April 1963," Box 11, Papers of Thomas E. Kurtz, MS-1144, Rauner Library (hereafter Kurtz Papers); John G. Kemeny and Thomas E. Kurtz, *Back to BASIC: The History, Corruption, and Future of the Language* (Reading, MA: Addison-Wesley, 1985).

2. Details drawn from John G. Kemeny and Thomas E. Kurtz, *The Dartmouth Time-Sharing Computing System: Final Report, June 1967*, Box 8, Stephen Garland Papers, ML-101, Rauner Library (hereafter Garland Papers), also available at http://eric.ed.gov/?id=ED024602; John G. Kemeny and Thomas E. Kurtz, "Dartmouth Time-Sharing," *Science* 162, no. 3850 (October 11, 1968): 223–228; Greg Dobbs (class of 1969), e-mail message to the author, September 4, 2013.

3. Jack Raymond, "President Asks 1.3 Billion for Missiles, Air Defense as Congress Reconvenes," *New York Times*, January 8, 1958, 1.

4. The Office of Charles & Ray Eames, *A Computer Perspective*, ed. Glen Fleck (Cambridge, MA: Harvard University Press, 1973), 162.

5. "704 Data Processing System," IBM Archives Online, http://www-03.ibm.com/ibm/history/exhibits/mainframe/mainframe_PP704.html.

6. For more on mainframes, see Martin Campbell-Kelly, William Aspray, Nathan Ensmenger, and Jeffrey R. Yost, *Computer: A History of the Information Machine*, 3rd ed. (Boulder, CO: Westview Press, 2014); Paul Ceruzzi, *A History of Modern Computing*, 2nd ed. (Cambridge, MA: MIT Press, 2003).

7. "704 Data Processing System."

8. General Electric Computer Department, "A Preliminary Proposal for Dartmouth College, October 15, 1962," Box 1, Kurtz Papers.

9. Kurtz, oral history interview.

10. Description of this type of programming in John G. Kemeny and Thomas E. Kurtz, "A Proposal for a College Computation Center," Bundle for Misc. Computer c. 1971–72 [bundle is labeled with these dates, even though it contains earlier materials], Box 11, Kurtz Papers.

11. Kemeny and Kurtz, "A Proposal for a College Computation Center"; Fernando J. Corbató, oral history interview, 1990, Charles Babbage Institute, retrieved from the University of Minnesota Digital Conservancy, http://purl.umn.edu/107230.

12. Harry M. Cantrell, report on visit to Dartmouth College, March 12, 1963, General Electric Proposal Binder, Box 1, Kurtz Papers, 1–2.

13. Ibid., 2.

14. William J. Miller, "A Visitor Looks at Dartmouth," booklet in Rauner Library, 2.

15. Special Correspondent, "Computers in the Home," *Nature* 212, no. 5058 (October 1966): 115.

16. Kurtz, oral history interview; Jere Daniell, conversation with the author, October 2011, Hanover, NH.

17. Kurtz, oral history interview.

18. John G. Kemeny, *Man and the Computer* (New York: Scribner, 1972).

19. Kurtz, Rieser, and Meck, "Application to the National Science Foundation," 42; John G. Kemeny, interview by A. Alexander Fanelli in Hanover, NH, during April–September 1984, DOH-2, Rauner Library.

20. Kurtz, Rieser, and Meck, "Application to the National Science Foundation," 42.

21. Ibid., 43–45a; John G. Kemeny, *Random Essays on Mathematics, Education, and Computers* (Englewood Cliffs, NJ: Prentice-Hall, 1964).

22. Kemeny, *Random Essays*, vii.

23. Kurtz, oral history interview, 7.

24. Ibid., 3.

25. Ibid.; Kurtz, Rieser, and Meck, "Application to the National Science Foundation," 41.

26. Kurtz, oral history interview.

27. Author's experience as an undergraduate mathematics major at Dartmouth, spending far too many hours in the Shower Towers. See also Scott Meacham and Joseph Mehling, *Dartmouth College: An Architectural Tour* (New York: Princeton Architectural Press, 2008).

28. Thomas Kurtz, Notes on Newsletters from Royal McBee and POOL, DOPE Folder, Box 3, Kurtz Papers; Kurtz, oral history interview, 13.

29. Kurtz, Rieser, and Meck, "Application to the National Science Foundation."

30. Product details from Royal Precision LGP-30 advertisement, Computer History Museum Early Companies Online Exhibition, http://www.computerhistory.org/revolution/early-computer-companies/5/116/1889.

31. Kurtz, Rieser, and Meck, "Application to the National Science Foundation"; Thomas E. Kurtz, "BASIC Session," in *History of Programming Languages*, ed. Richard L. Wexelblat (New York: Association for Computing Machinery [ACM], 1981), 515–549.

32. Various LGP-30 Users' Group Newsletters, Box 3, Kurtz Papers.

33. Kurtz, Rieser, and Meck, "Application to the National Science Foundation"; Kurtz, "BASIC Session."

34. Kurtz, Rieser, and Meck, "Application to the National Science Foundation."

35. Ibid., 10.

36. "Dartmouth Oversimplified Programming Experiment, May 1962," DOPE Folder, Box 3, Kurtz Papers.

37. Ibid., 1.

38. Kurtz, Rieser, and Meck, "Application to the National Science Foundation," 11.

39. Kemeny and Kurtz, "A Proposal for a College Computation Center."

40. Ibid., 4.

41. Ibid.

42. Kurtz corresponded with Fernando Corbató at MIT about time-sharing in November 1961. Fernando Corbató to Thomas E. Kurtz, November 13, 1961, Bundle for Misc. Computer c. 1971–72, Box 11, Kurtz Papers.

43. The earliest contemporaneous articulation of these ideas appears in Kurtz, Rieser, and Meck, "Application to the National Science Foundation" from April 1963. Kemeny presented these ideas to the Dartmouth College Board of Trustees in April 1963, per John G. Kemeny, "A Computing Center at a Liberal Arts College," in *Random Essays on Mathematics, Education and Computers* (Englewood Cliffs, NJ: Prentice-Hall, 1964), 157–163. However, Kemeny and Kurtz concur in later recollections that these ideas were Kurtz's and that he presented them to Kemeny. See, for example, Transcript of 1974 National Computer Conference Pioneer Day Session, Box 1, Papers of Sidney Marshall, MS-1054, Rauner Library (hereafter Marshall Papers). See also Kemeny and Kurtz, *Back to BASIC.*

44. Kurtz, oral history interview; Corbató, oral history interview.

45. Corbató, oral history interview. Kurtz's papers include an undated article about time-sharing by McCarthy, J. C. R. Licklider, and others, as well as a February 1961 paper by Corbató and others, Bundle for Misc. Computer c. 1971–72, Box 11, Kurtz Papers.

46. Fernando J. Corbató, Marjorie Merwin-Daggett, and Robert C. Daley, "An Experimental Time-Sharing System," in *Proceedings of the May 1–3, 1962, Spring Joint Computer Conference*, AIEE-IRE '62 (Spring) (New York: ACM, 1962), 335–344, doi:10.1145/1460833.1460871.

47. Ibid.

48. Kurtz, "BASIC Session," 519.

49. Fernando Corbató to Thomas E. Kurtz, November 13, 1961, Bundle for Misc. Computer c. 1971–72, Box 11, Kurtz Papers; Thomas E. Kurtz, notes on Time Sharing Work, Bundle for Misc. Computer c. 1971–72, Box 11, Kurtz Papers.

50. Corbató, oral history interview; Arthur L. Norberg, Judy E. O'Neill, and Kerry J. Freedman, *Transforming Computer Technology: Information Processing for the Pentagon, 1962–1986* (Baltimore: Johns Hopkins University Press, 1996); M. Mitchell Waldrop, *The Dream Machine: J. C. R. Licklider and the Revolution That Made Computing Personal* (New York: Penguin Books, 2002); Steven Levy, *Hackers: Heroes of the Computer Revolution* (Sebastopol, CA: O'Reilly Media, 2010).

51. Corbató, Merwin-Daggett, and Daley, "An Experimental Time-Sharing System," 337.

52. Kurtz, Rieser, and Meck, "Application to the National Science Foundation"; Kemeny, "A Computing Center at a Liberal Arts College."

53. Kemeny, "A Computing Center at a Liberal Arts College," 162.

54. Kemeny and Kurtz, *Back to BASIC*, 4. Kurtz also conveyed this in the memo labeled 1962, Folder UP 128 1:1, Box 1, Kurtz Papers.

55. "A Data Processing System Report for Dartmouth College," Box 1, Kurtz Papers.

56. Kemeny and Kurtz had planned on the GE-235 for their system from nearly the beginning of their negotiations with GE, but when they learned it would not be available until autumn 1964, they agreed to accept the GE-225 in the interim, based on Kurtz, Rieser, and Meck, "Application to the National Science Foundation."

57. Kurtz, Rieser, and Meck, "Application to the National Science Foundation"; John G. Kemeny and Thomas E. Kurtz, "Time Sharing Project Memo No. 2—Outline of the Executive System," Folder 265—Early Papers, Box 4248, Records of Dartmouth College Computing Services (hereafter RDCCS), DA-181, Rauner Library; Anthony Knapp, e-mail messages to the author, July 18 and 21, 2013.

58. Corbató, oral history interview; Corbató, Merwin-Daggett, and Daley, "An Experimental Time-Sharing System."

59. Knapp, e-mail message to the author, July 21, 2013.

60. John G. Kemeny, "The GE-Dartmouth Computer Partnership," DTSS History Birth 1967–70 Folder 2, Box 4, Garland Papers. Corroborated by Anthony Knapp, e-mail messages to the author, July 18 and 21, 2013.

61. Kemeny, "The GE-Dartmouth Computer Partnership," 2.

62. Ibid.; Thomas Kurtz, Memo on 1962, Folder UP 128 1:1, Box 1, Kurtz Papers.

63. Thomas Kurtz and Anthony Knapp, "Master Executive Supervisory System (MESS)," Dartmouth College Computation Center, 1962. Kurtz referenced this document in Kurtz, "BASIC Session." Knapp confirmed its existence and purpose in an e-mail message to the author on July 21, 2013.

64. Anthony Knapp, e-mail message to the author, July 18, 2013.

65. General Electric Computer Department, "A Preliminary Proposal for Dartmouth College, October 15, 1962," Box 1, Kurtz Papers.

66. Ibid., V-1.

67. John G. Kemeny to Dean Myron Tribus, April 2, 1963, GE Proposal Binder, Box 1, Kurtz Papers. Kemeny reported, "We are hopefully reaching the final stage of negotiations with GE." The commitment to GE equipment was also expressed in Kurtz, Rieser, and Meck, "Application to the National Science Foundation."

68. Kemeny, *Random Essays.*

69. M. Mitchell Waldrop, *The Dream Machine: J. C. R. Licklider and the Revolution That Made Computing Personal* (New York: Penguin Books, 2002); Thierry Bardini, *Bootstrapping: Douglas Engelbart, Coevolution, and the Origins of Personal Computing* (Stanford, CA: Stanford University Press, 2000).

70. Kemeny, *Random Essays.*

71. Kurtz, "BASIC Session," 519–520.

72. Kurtz, "BASIC Session."

73. Kemeny, "A Computing Center at a Liberal Arts College."

74. Kurtz, oral history interview, 25.

75. Kurtz, Rieser, and Meck, "Application to the National Science Foundation," cover page.

76. Ibid., 5.

77. Notification of award from John G. Kemeny and Thomas E. Kurtz, "Dartmouth Time-Sharing," *Science* 162, no. 3850 (October 11, 1968): 223–228. Contract details from Kurtz, Memo on 1962, Folder UP 128 1:1, Box 1, Kurtz Papers.

78. "Dartmouth Time Sharing Important Historical Dates," DTSS Prehistory GE-235 Folder, Box 4, Garland Papers; John G. Kemeny, *NSF Grant GE-3864 in Support of Experiments in Undergraduate Instruction on Computing: Report of First Year Progress, September 1, 1964*, Secondary School Folder, Box 4247, RDCCS, DA-181, Rauner Library; Kemeny and Kurtz, *The Dartmouth Time-Sharing Computing System.*

79. Transcript of 1974 National Computer Conference Pioneer Day Session; Kemeny and Kurtz, *Back to BASIC.*

80. Transcript of 1974 National Computer Conference Pioneer Day Session; Kemeny and Kurtz, *Back to BASIC.* A note on the acronym: The "Beginners" in BASIC has appeared as "Beginner's" or "Beginners'" (apostrophe before the "s" or apostrophe after the "s"). For example, the commemorative edition reprint of the draft BASIC manual from May 1964, available at http://www.dartmouth.edu/basicfifty/basicmanual_1964.pdf, uses "Beginners'"; the BASIC manual from October 1964, published by the Dartmouth College

Computation Center and located in the Dartmouth BASIC Folder, Box 8, Garland Papers, uses "Beginner's" (on page 2). I have used "Beginners'" because it has been more frequently used and because it implies multiple beginners learning the language.

81. Kurtz reported that Kemeny developed the first BASIC compiler during the summer of 1963 in his Memo on 1962, Folder UP 128 1:1, Box 1, Kurtz Papers. His memo on the time-sharing project dated November 1963 confirmed that the compiler had already been written: Thomas Kurtz, "Time Sharing Project Memo No. 1," Folder 265—Early Papers, Box 4248, RDCCS, DA-181, Rauner Library. See also John G. Kemeny, Peter Lawes, and William Zani, "Time Sharing Project Memo No. 3—Outline of BASIC," Folder 265—Early Papers, Box 4248, RDCCS, DA-181, Rauner Library.

82. Kurtz, "Time Sharing Project Memo No. 1."

83. Ibid., 1-1.

84. Ibid.

85. Ibid., 1-2; emphasis in original.

86. Kemeny and Kurtz, "Time Sharing Project Memo No. 2—Outline of the Executive System," Folder 265—Early Papers, Box 4248, RDCCS, DA-181, Rauner Library.

87. Kemeny described this at Pioneer Day, Transcript of 1974 National Computer Conference Pioneer Day Session, and in Kemeny and Kurtz, *Back to BASIC.*

88. For a classic examination of technological systems, see Thomas Parke Hughes, *Networks of Power: Electrification in Western Society, 1880–1930* (Baltimore: Johns Hopkins University Press, 1993).

89. The evolution of this requirement from a one-hour computing laboratory to a computing course embedded in required mathematics courses can be traced in the documents dated October 1963, Bundle for Misc. Computer c. 1971–72, Box 11, Kurtz Papers.

90. Kemeny, "A Computing Center at a Liberal Arts College," 157; Kemeny and Kurtz, "Dartmouth Time-Sharing" (*Science*).

91. Richard H. Crowell, Peter Lawes, Robert Z. Norman, and John Warren, "Time Sharing Project Memo No. 4—Mathematics Course Computing Supplement," Folder 265—Early Papers, Box 4248, RDCCS, DA-181, Rauner Library.

92. Kemeny and Kurtz, *Back to BASIC*, 5.

93. The GE manuals have been preserved and are available online through a website commemorating the DTSS. See, for example, General Electric, "GE-235 Central Processor Reference Manual, March 1964," http://dtss.dartmouth.edu/scans/235%20Manuals/235%20CPU%20Reference.pdf.

94. Ibid.; General Electric, "GE-235 Mass Random Access Data Storage," http://dtss.dartmouth.edu/scans/235%20Manuals/DiskManual/p-viii-1.tif.

95. General Electric, "GE-235 Central Processor Reference Manual, March 1964," VIII-4–VIII-30. The Dartmouth group worked with a GE-225 machine from February through September 1964, then upgraded to the similar GE-235 in September 1964.

96. John G. Kemeny, "Datanet-30 Output," Folder 265—Early Papers, Box 4248, RDCCS, DA-181, Rauner Library.

97. Kemeny, *NSF Grant GE-3864*.

98. Ibid.; McGeachie in Transcript of 1974 National Computer Conference Pioneer Day Session.

99. Transcript of 1974 National Computer Conference Pioneer Day Session.

100. Kemeny in Transcript of 1974 National Computer Conference Pioneer Day Session, 5.

101. McGeachie in Transcript of 1974 National Computer Conference Pioneer Day Session, 5.

102. Kemeny, *NSF Grant GE-3864*.

103. Kemeny placed the start of time-sharing sometime in May 1964 in his September 1964 progress report, *NSF Grant GE-3864*. The date of May 1, 1964, was discussed at the Pioneer Day Session, Transcript of 1974 National Computer Conference Pioneer Day Session. Kurtz ultimately acknowledged that the May 1, 1964, at 4:00 a.m. date was "a pretty good myth" in Kurtz, oral history interview, 22.

104. "Attention: Time-Sharing Users, July 30, 1964," in Folder DTSS Prehistory GE-235, Box 4, Garland Papers.

105. Ibid., 1; emphasis in original.

106. Ibid., 2.

107. Estimated student numbers in Computer Project, Freshman Committee, Meeting Notes for Wednesday, October 30, 1963, Bundle for Misc. Computer c. 1971–72, Box 11, Kurtz Papers; Crowell et al., "Time Sharing Project Memo No. 4."

108. Crowell et al., "Time Sharing Project Memo No. 4"; Computer Project, Freshman Committee, Meeting Notes for Wednesday, October 9, 1963, Bundle for Misc. Computer c. 1971–72, Box 11, Kurtz Papers.

109. Kemeny, *NSF Grant GE-3864*.

110. Ibid., 6.

111. Ibid., 7.

112. Ibid.

113. [No author listed, but based on contents and surrounding material, the author is almost certainly Kurtz], "Course Content Improvement Project Informal Report," Secondary School Folder, Box 4247, RDCCS, DA-181, Rauner Library.

114. Kurtz, Rieser, and Meck, "Application to the National Science Foundation."

115. [Kurtz], "Course Content Improvement Project Informal Report."

116. Ibid., 3.

117. Overheads for Fall Term Course, November 1964, Folder UP 128 11:1, Box 11, Kurtz Papers.

118. Dartmouth College Computation Center, "BASIC: A Manual for BASIC, the Elementary Algebraic Language Designed for Use with the Dartmouth Time Sharing System" (Hanover, NH: Trustees of Dartmouth College, [October 1,] 1964), 15, Dartmouth BASIC Folder, Box 8, Garland Papers; Jim Lawrie (class of 1968), e-mail message to the author, August 28, 2013.

119. Dartmouth College Computation Center, "BASIC October 1964."

120. Ibid.; "Attention: Time-Sharing Users, July 30, 1964."

2 Making a Macho Computing Culture

1. Transcript of 1974 National Computer Conference Pioneer Day Session, 12, Box 1, Papers of Sidney Marshall, MS-1054, Dartmouth College Rauner Special Collections Library (hereafter Rauner Library).

2. The gift of $500,000 is cited in a clipping from the January 1965 *Dartmouth Alumni Magazine.* Peter Kiewit details from the brochure *The Future Impact of Computers: A Conference to Dedicate the Kiewit Computation Center at Dartmouth College* and the article "The Man behind the Name," *Dartmouth Alumni Magazine,* January 1967, 15. All are in Information File for the Computation Center, Rauner Library.

3. Stuart H. Loory, "Dartmouth Expands Computer Facility," *New York Times,* December 4, 1966; and Stuart H. Loory, "Computer Hookup to Home Foreseen," *New York Times,* December 5, 1966. Clippings from *The Dartmouth* and the *Dartmouth Alumni Magazine* are in Information File for the Computation Center, Rauner Library.

4. From the remarks by Professor John G. Kemeny, chair of Dartmouth's Department of Mathematics, at the dedication, in the *Dartmouth Alumni Magazine,* January 1967, 18, in Information File for the Computation Center, Rauner Library.

5. Brochure, *The Kiewit Computation Center & the Dartmouth Time-Sharing System,* in Information File for Time-Sharing, Rauner Library.

6. Loory, "Dartmouth Expands Computer Facility."

7. Ibid.

8. The Dartmouth Computation Center FACT SHEET, September 22, 1966, in Information File for the Computation Center, Rauner Library.

9. *Kiewit Comments* 3, no. 4 (April 20, 1969), Rauner Library. Please note that all citations to *Kiewit Comments* are drawn from the Rauner Library collection so subsequent notes cite only individual newsletter details.

10. Kiewit Technical Memoranda 10, DTSS Library Programs, Box 7, Stephen Garland Papers, ML-101, Rauner Library (hereafter Garland Papers). Please note that the full versions of these program names include three asterisks, for example, BANDIT***, FTBALL***, or GRIDIRON***, but I have eliminated the asterisks for ease of comprehension.

11. Kurtz in *Kiewit Comments* 1, no. 9 (December 18, 1967): 2. *Kiewit Comments* 1, no. 7 (October 11, 1967): 6 announced to users: "URGENT!! . . . PLEASE UNSAVE ALL OF YOUR INACTIVE PROGRAMS"; capitalization and emphasis in original.

12. Quote and paper tape in *Kiewit Comments* 2, no. 8 (October 23, 1968): 1; punched cards in *Kiewit Comments* 2, no. 5 (May 24, 1968): 4.

13. *Kiewit Comments* 2, no. 1 (January 15, 1968): 2.

14. *Kiewit Comments* 1, no. 3 (March 22, 1967): 3.

15. *Kiewit Comments* 2, no. 4 (April 17, 1968): 2.

16. *Kiewit Comments* 4, no. 6 (September 28, 1970).

17. John G. Kemeny, *Random Essays on Mathematics, Education and Computers* (Englewood Cliffs, NJ: Prentice-Hall, 1964); John G. Kemeny, *Man and the Computer* (New York: Scribner, 1972); John G. Kemeny, "The Question of

Networks: What Kind and Why?," *EDUCOM: Bulletin of the Interuniversity Communications Council* 8, no. 2 (Summer 1973): 18–21.

18. Special Correspondent, "Computers in the Home," *Nature* 212, no. 5058 (October 1966): 115.

19. "Playing a game of football" on the time-sharing system is mentioned in the brochure *The Kiewit Computation Center & The Dartmouth Time-Sharing System*, 4. *Kiewit Comments* 3, no. 7 (November 4, 1969) and *Kiewit Comments* 4, no. 6 (September 28, 1970) discuss FOOTBALL. Kemeny also mentioned it in his prominent lecture "Large Time-Sharing Networks," part of a series sponsored by the Johns Hopkins University and the Brookings Institution during 1969–1970. Manuscript, with football on pages 11–12, in DTSS Info Publications folder, Box 4, Garland Papers.

20. "Students' Studies Geared to Time-Shared Computer," *Electrical World*, October 1968, in Folder 1964–67, Box 4242, Records of Dartmouth College Computing Services (hereafter RDCCS), DA-181, Rauner Library.

21. Transcript of 1974 National Computer Conference Pioneer Day Session, 8.

22. Ibid., 1.

23. For a cogent examination of policing Cold War heteronormative gender roles, see Alexandra Minna Stern, *Eugenic Nation: Faults and Frontiers of Better Breeding in Modern America* (Berkeley: University of California Press, 2015), particularly chapter 6. See also Joy Lisi Rankin, "Eugenics: Policing Everything," *Lady Science* 33 (June 2017), https://www.ladyscience.com/archive/sciencegender fascismp2. The literature on postwar American gender, sexuality, and family is extensive. A few examples include Elaine Tyler May, *Homeward Bound: American Families in the Cold War Era*, fully revised and updated 20th anniversary edition (New York: Basic Books, 2008); Joanne Meyerowitz, ed., *Not June Cleaver: Women and Gender in Postwar America, 1945–1960* (Philadelphia: Temple University Press, 1994); Beth Bailey, *Sex in the Heartland* (Cambridge, MA: Harvard University Press, 2002); David Johnson, *The Lavender Scare: The Cold War Persecution of Gays and Lesbians in the Federal Government* (Chicago: University of Chicago Press, 2004); Margot Canaday, *The Straight State: Sexuality and Citizenship in Twentieth-Century America* (Princeton, NJ: Princeton University Press, 2009).

24. Francis Marzoni, e-mail message to the author, December 2, 2013.

25. Medford Cashion III, e-mail message to the author, December 3, 2013.

26. David Ziegler, e-mail message to the author, December 18, 2013.

27. *Kiewit Comments* 3, no. 7 (November 4, 1969): 6–7.

28. For entry into this extensive literature, see Justine Cassell and Henry Jenkins, eds., *From Barbie® to Mortal Kombat: Gender and Computer Games* (Cambridge, MA: MIT Press, 2000); Yasmin B. Kafai, Carrie Heeter, Jill Denner, and Jennifer Y. Sun, eds., *Beyond Barbie® and Mortal Kombat: New Perspectives on Gender and Gaming* (Cambridge, MA: MIT Press, 2008); danah boyd, *It's Complicated: The Social Lives of Networked Teens* (New Haven, CT: Yale University Press, 2015).

29. *Kiewit Comments* 4, no. 5 (June 17, 1970): 10.

30. *Kiewit Comments* 5, no. 1 (January 11, 1971): 10.

31. For an overview of the history and historiography of computing and gender, see Joy Lisi Rankin, "Queens of Code," *Lady Science* no. 9 (June 2015), https://www.ladyscience.com/queens-of-code.

32. Thomas C. Jepsen, *My Sisters Telegraphic: Women in the Telegraph Office, 1846–1950* (Athens: Ohio University Press, 2000).

33. Author's e-mail correspondence with Dartmouth alumni; see specific citations above.

34. Janet Price, "The Use of Conjunctive and Disjunctive Models in Impression Formation" (PhD diss., Dartmouth College, 1971); Janet Price, ["Autobiography"], *The Lathrop Nor'Easter*, series II: vol. 1–4 (Summer 2016): 11, http://lathrop.kendal.org/wp-content/uploads/sites/3/2017/04/NorEaster Summer2016.pdf.

35. Nancy Broadhead, "Consulting Is Not a Matter of Facts," in *SIGUCCS '81: Proceedings of the 9th Annual ACM SIGUCCS Conference on User Services* (New York: ACM, October 1981); Nancy Broadhead, "The Unique Advantages of the Macintosh," in *SIGUCCS '85: Proceedings of the 13th Annual ACM SIGUCCS Conference on User Services: Pulling It All Together* (New York: ACM, September 1985).

36. *Kiewit Comments* 2, no. 6 (August 19, 1968); *Kiewit Comments* 3, no. 7 (November 4, 1969).

37. *Kiewit Comments* 2, no. 7 (September 20, 1968).

38. *Kiewit Comments* 3, no. 6 (September 20, 1969): 10.

39. Kiewit Computation Center, *BASIC* [prepared by the staff of the Dartmouth College Kiewit Computation Center and revised by Diane Mather] (Hanover, NH, 1970), Rauner Library; Kiewit Computation Center, *DTSS Library Programs. By: Diane Mather* [and] *Len Smith* (Hanover, NH, 1973), Rauner Library.

40. Price and DeNood in *Kiewit Comments* 2, no. 1 (January 15, 1968): 4. In *Kiewit Comments* 1, no. 2 (February 13, 1967): 5, "The new editor . . . is (Mrs.) Lois D. Woodard." In *Kiewit Comments* 1, no. 5 (June 19, 1967): 2, "A new addition to the Computation Center staff is Mrs. Frances (Bonnie) Clark." In *Kiewit Comments* 2, no. 2 (February 19, 1968): 2, users are advised to call "Mrs. Janet Price . . . Mrs. Nancy Broadhead." *Kiewit Comments* 2, no. 7 (September 20, 1968): 2, introduced "Jack Nevison . . . Mrs. Jean Danver . . . Roger Blake . . . Mr. William S. Shaw . . . Mrs. Ruth Bogart." *Kiewit Comments* 3, no. 3 (March 21, 1969): 2–3, bid farewell to "Mrs. Susan Merrow" and "Miss Linda Caldwell," while *Kiewit Comments* 3, no. 5 (June 1, 1969): 8, welcomed "Miss Diane Sonneborn" and "Mrs. Donna Zuba." For more on women computers, see Margot Lee Shetterly, *Hidden Figures: The American Dream and the Untold Story of the Black Women Mathematicians Who Helped Win the Space Race* (New York: HarperCollins, 2016).

41. *Kiewit Comments* 2, no. 5 (May 24, 1968): 2.

42. *Kiewit Comments* 2, no. 8 (October 23, 1968): 4; emphasis in original.

43. *Kiewit Comments* 3, no. 7 (November 4, 1969): 8. Other examples include *Kiewit Comments* 2, no. 7 (September 20, 1968) reporting on the arrival of social sciences programmer Mrs. Ruth Bogart and mentioning that her husband, Ken, was an assistant professor of mathematics. *Kiewit Comments* 3, no. 3 (March 21, 1969) noted that when Mrs. Susan Merrow left Kiewit, she was going to tour Europe with her husband, Ed. *Kiewit Comments* 4, no. 5 (June 17, 1970)

reported that Diane Hills was leaving because her husband, Jerry, would be attending graduate school in Boston. I will note that *Kiewit Comments* 4, no. 4 (April 8, 1970) welcomed the daughter of staffer Sammy Karumba, and *Kiewit Comments* 4, no. 6 (September 28, 1970) described the wife (Halina Mlotkowski) of staffer Sidney Marshall; however, there are only a handful of these mentions of the family life of the male staffers compared with the more frequent discussions of the husbands and children of the female staffers during these formative years of the campus computing culture.

44. *Kiewit Comments* 2, no. 10 (December 20, 1968): 7.

45. 7 / 26 / 72 Memo on DTSS—the Dartmouth Time-Sharing System from RFH [Robert F. Hargraves] and JSM [John S. McGeachie], Information File for Time-Sharing, Rauner Library.

46. Transcript of 1974 National Computer Conference Pioneer Day Session.

47. For an analysis of the history and masculinity of the "computer nerd," see Nathan Ensmenger, "'Beards, Sandals, and Other Signs of Rugged Individualism': Masculine Culture within the Computing Professions," *Osiris* 30 (2015): 38–65.

48. Thomas E. Byrne and Christine W. Thompson, eds., *Computing at Dartmouth 1973–1976: A Report on the Activities of the Kiewit Computation Center*, 9, Box 9, Garland Papers.

49. Ibid. I examined photographs for captions and facial and clothing markers that were characteristically gendered masculine or feminine. I identified approximately fifty-five men and eight women. Although these totals may vary slightly, the overall proportion of men to women in the photos is the point, and it is dramatically skewed toward the men.

50. Ibid., 18.

51. William J. Miller, "A Visitor Looks at Dartmouth," booklet in Rauner Library.

52. Ibid., 22.

53. Ibid., 23.

54. Richard L. Zweigenhaft and G. William Domhoff, *Blacks in the White Elite: Will the Progress Continue* (Lanham, MD: Rowman & Littlefield, 2003); Shereen Meraji, "Fifty Years Later, 'A Better Chance' Trains Young Scholars," National Public Radio (NPR), June 9, 2013, http://www.npr.org/sections/codeswitch/2013/06/09/184798293/fifty-years-later-a-better-chance-trains-young-scholars; Bill Platt, "Forty Years On: The Changing Face of Dartmouth," *Dartmouth News*, June 17, 2013, https://news.dartmouth.edu/news/2013/06/forty-years-changing-face-dartmouth.

55. Miller, "A Visitor Looks at Dartmouth," 23–24.

56. Platt, "Forty Years On."

57. Ibid.

58. *Kiewit Comments* 2, no. 7 (September 20, 1968): 8.

59. Miller, "A Visitor Looks at Dartmouth," 23.

60. *Kiewit Comments* 4, no. 4 (April 8, 1970): 11.

61. *Kiewit Comments* 4, no. 5 (June 17, 1970): 1.

62. "The DTSS as a Regional Resource" section (no page numbers) in *The Kiewit Computation Center & the Dartmouth Time-Sharing System* (brochure) [n.d., c.1971] in the Information File for Time Sharing, Rauner Library.

63. *Kiewit Comments* 4, no. 6 (September 28, 1970): 7.

64. *Final Report: Time-Sharing Computer Applications in Undergraduate Anthropology at Dartmouth College*, 4, in Folder: DA 29 (7841) 2 of 6, labeled: Time-Sharing Computer / Kiewit, Rauner Library.

65. Ibid.

66. For entry into the literature on race and computing, see Safiya U. Noble, *Algorithms of Oppression: How Search Engines Reinforce Racism* (New York: New York University Press, 2018); Lisa Nakamura and Peter A. Chow-White, eds., *Race after the Internet* (New York: Routledge, 2012); Lisa Nakamura, *Digitizing Race: Visual Cultures of the Internet* (Minneapolis: University of Minnesota Press, 2007).

67. Thomas E. Kurtz, Leonard M. Rieser, and John F. Meck, "Application to the National Science Foundation Mathematics Division for the Establishment and Support of the Dartmouth Computation Center, 22 April 1963," 31–32, Papers of Thomas E. Kurtz, MS-1144, Rauner Library.

68. "The DTSS as a Regional Resource" section.

69. Ibid.

70. *Kiewit Comments* 2, no. 3 (March 15, 1968): cover.

71. *The Kiewit Computation Center & the Dartmouth Time-Sharing System;* The Kiewit Computation Center, *Biennial Report 1969–1971*, 17, in Box 9, Garland Papers.

72. The Kiewit Computation Center, *Biennial Report 1971–1973*, 20, in Box 9, Garland Papers.

73. *Kiewit Comments* 2, no. 1 (January 15, 1968) and *Kiewit Comments* 4, no. 4 (April 8, 1970).

74. Janet Price from Washington in *Kiewit Comments* 2, no. 1 (January 15, 1968); Jean Danver from Massachusetts and Ruth Bogart from California in *Kiewit Comments* 2, no. 7 (September 20, 1968); Diane Mather from New York in *Kiewit Comments* 3, no. 6 (September 20, 1969).

75. Hatfield students in *Kiewit Comments* 2, no. 4 (April 17, 1968) and *Kiewit Comments* 4, no. 4 (April 8, 1970); Jean Louis Leonhardt from Lyon in *Kiewit Comments* 3, no. 6 (September 20, 1969) and *Kiewit Comments* 4, no. 6 (September 28, 1970).

76. Pillsbury in *Kiewit Comments* 1, no. 5 (June 19, 1967), and in Kristin Serum, "Computer System Installed in Schools," *Minneapolis Star*, January 19, 1967, from Folder 1964–67, Box 4242, RDCCS, DA-181, Rauner Library. Tuskegee in *Kiewit Comments* 2, no. 10 (December 20, 1968), *Kiewit Comments* 4, no. 4 (April 8, 1970), and in 29 September 1969 letter from Dr. D. L. Shell, Manager (GE) to Professor T. Kurtz re: Visit to Dartmouth Last Week, Box 4, Garland Papers.

77. *Kiewit Comments* 3, no. 7 (November 4, 1969): 9.

78. Details drawn from Kemeny and Kurtz, *The Dartmouth Time-Sharing Computing System. Final Report*, Box 8, Garland Papers; John G. Kemeny and Thomas E. Kurtz, "Dartmouth Time-Sharing," *Science* 162, no. 3850 (October 11, 1968): 223–228; Greg Dobbs (class of 1969), e-mail message to author, September 4, 2013.

3 Back to BASICs

1. LeBrun refers to BASIC as a "lingua franca" in Marc LeBrun, "A Polemic: Tilting at Windmills, or What's Wrong with BASIC?," *People's Computer Company* 1, no. 2 (December 1972): 5, Stanford University Digital Archives, https://purl.stanford.edu/cj445qq4021. Note that all subsequent references to the *People's Computer Company* refer to the Stanford University digital collection. The title of this chapter recalls the book by John G. Kemeny and Thomas E. Kurtz, *Back to BASIC: The History, Corruption, and Future of the Language* (Reading, MA: Addison-Wesley, 1985).

2. Ksenia Tatarchenko analyzed the negotiations that forged English as the "lingua franca" of computer science in "Not Lost in Translation," a paper presented in 2010 at the Software for Europe workshop, cited in the introduction to *Hacking Europe: From Computer Culture to Demoscenes*, ed. Gerard Alberts and Ruth Oldenziel (London: Springer-Verlag, 2014), 11.

3. For more on the history of software, see Martin Campbell-Kelly, *From Airline Reservations to Sonic the Hedgehog: A History of the Software Industry* (Cambridge, MA: MIT Press, 2003); Michael S. Mahoney, *Histories of Computing*, ed. Thomas Haigh (Cambridge, MA: Harvard University Press, 2011).

4. "Computer Jumps to Ski Conclusions," *New York Times*, February 5, 1967.

5. *Kiewit Comments* 1, no. 2 (February 13, 1967): 5, Dartmouth College Rauner Special Collections Library (hereafter Rauner Library). Note that all subsequent references to *Kiewit Comments* refer to the Rauner Library collection.

6. United States President's Science Advisory Committee, Panel on Computers in Education (John R. Pierce et al.), *Computers in Higher Education: Report of the President's Science Advisory Committee* (Washington, DC: The White House, February 1967).

7. Detail on copies sold from Steven Levy, *Hackers: Heroes of the Computer Revolution* (Sebastopol, CA: O'Reilly Media, 2010), 168; and from Bob Johnstone, *Never Mind the Laptops: Kids, Computers, and the Transformation of Learning* (New York: iUniverse, 2003), 66.

8. "Huntington," *People's Computer Company* 1, no. 1 (October 1972): 3.

9. Thomas E. Kurtz, Leonard M. Rieser, and John F. Meck, "Application to the National Science Foundation Mathematics Division for the Establishment and Support of the Dartmouth Computation Center, 22 April 1963," 10, Box 11, Papers of Thomas E. Kurtz, MS-1144, Rauner Library.

10. John G. Kemeny, *NSF Grant GE-3864 in Support of Experiments in Undergraduate Instruction on Computing: Report of First Year Progress, September 1, 1964*, Secondary School Folder, Box 4247, Records of Dartmouth College Computing Services (hereafter RDCCS), DA-181, Rauner Library.

11. Ibid.

12. Thomas E. Kurtz, interview by Daniel Daily in Hanover, NH, on June 20, 2002, and July 2, 2002, DOH-44, Rauner Library, 31.

13. Kemeny, *NSF Grant GE-3864*, 5.

14. Ibid.

15. Ibid. Kemeny's exact phrase was "as far as the user is concerned, the central processor talks BASIC."

16. Robert E. Smith and Dora E. Johnson, *FORTRAN Autotester* (New York: John Wiley & Sons, 1962), title page.

17. Ibid., 38.

18. For example, Kemeny's 1964 first-year progress report on the NSF GE-3864 grant claimed that BASIC could be mastered in three hours, and Kemeny and Kurtz's 1967 final report on the NSF GE-3864 grant explained that first-year students learned BASIC in two hour-long lectures.

19. John G. Kemeny and Thomas E. Kurtz, *BASIC Programming* (New York: John Wiley & Sons, 1967), 3–5.

20. Contents of this paragraph drawn from Dartmouth College Computation Center, "BASIC: A Manual for BASIC, the Elementary Algebraic Language Designed for Use with the Dartmouth Time Sharing System" (Hanover, NH: Trustees of Dartmouth College, [October 1,] 1964), Dartmouth BASIC Folder, Box 8, Stephen Garland Papers (hereafter Garland Papers), ML-101, Rauner Library.

21. Kemeny, *NSF Grant GE-3864.*

22. Kurtz, Rieser, and Meck, "Application to the National Science Foundation," 32.

23. John G. Kemeny and Thomas E. Kurtz, *The Dartmouth Time-Sharing Computing System. Final Report*, 25, Box 8, Garland Papers, also available at http://eric.ed.gov/?id=ED024602; Gordon D. Goldstein, Laura A. Repass, and Barbara J. Walker, *Digital Computer Newsletter* 20, no. 1 (Arlington, VA: Office of Naval Research, January 1968), http://www.dtic.mil/docs/citations/AD 0694655.

24. *Kiewit Comments* 1, no. 3 (March 22, 1967): 4.

25. *Kiewit Comments* 1, no. 6 (August 9, 1967): 5; project descriptions of Hanover High School students also in Special Student Applications, in Secondary School Folder, Box 4247, RDCCS, DA-181, Rauner Library.

26. Kemeny and Kurtz, *The Dartmouth Time-Sharing Computing System;* Goldstein, Repass, and Walker, *Digital Computer Newsletter*, 11.

27. Transcript of 1974 National Computer Conference Pioneer Day Session, Box 1, Papers of Sidney Marshall, MS-1054, Rauner Library.

28. Ibid., 13.

29. Ibid.

30. Francis Marzoni, e-mail message to author, December 2, 2013.

31. Special Correspondent, "Computers in the Home," *Nature* 212, no. 5058 (October 1966): 115–116, doi:10.1038/212115f0.

32. Kemeny and Kurtz, *The Dartmouth Time-Sharing Computing System*, 24–26.

33. Kemeny and Kurtz, *The Dartmouth Time-Sharing Computing System.*

34. Ibid.; Goldstein, Repass, and Walker, *Digital Computer Newsletter.*

35. Kemeny and Kurtz, *The Dartmouth Time-Sharing Computing System.*

36. Thomas E. Kurtz, *Demonstration and Experimentation in Computer Training and Use in Secondary Schools: Interim Report, Activities and Accomplishments of the First Year*, October 1968, Dartmouth College History Collection, Rauner Library.

37. Claudia Goldin, Lawrence F. Katz, and Ilyana Kuziemko, "The Homecoming of American College Women: The Reversal of the Gender Gap in College," *Journal of Economic Perspectives* 20 (2006): 133–156.

38. Time Sharing Project Memo No. 4, "Mathematics Course Computing Supplement" by Richard H. Crowell, Peter Lawes, Robert Z. Norman, and John Warren, December 16, 1963, in Box 8, Garland Papers, and Box 4248, RDCCS, Rauner Library; Kemeny, *NSF Grant GE-3864*.

39. James F. Bowring, "A Time-Sharing Computer at Exeter," April 1966, clipping in Folder DA 29 (2841) 4 of 6, labeled: Articles Time-Sharing, Rauner Library.

40. Ibid.

41. Ibid.

42. Jean H. Danver, *Demonstration and Experimentation in Computer Training and Use in Secondary Schools, Interim Report II*, December 1969, Dartmouth College History Collection, Rauner Library.

43. Kurtz, *Demonstration and Experimentation in Computer Training and Use in Secondary Schools: Interim Report.*

44. Dickey quoted in William J. Miller, "A Visitor Looks at Dartmouth," 21, booklet in Rauner Library.

45. Kurtz, *Demonstration and Experimentation in Computer Training and Use in Secondary Schools: Interim Report*, 30. Although Kurtz is listed as the report's author, he noted in his foreword that project coordinator John Nevison was "the principle [*sic*] author of the main body of this report."

46. Ibid., 9. Dartmouth's motto is *Vox clamantis in deserto.*

47. Ibid., 3.

48. Ibid., 11.

49. Ibid. Keith K. Thompson and Ann Waterhouse, "The Time-Shared Computer: A Teaching Tool," *NASSP Bulletin* 54, no. 343 (February 1, 1970): 91–98, doi:10.1177/019263657005434312.

50. Goldstein, Repass, and Walker, *Digital Computer Newsletter.*

51. Kurtz, *Demonstration and Experimentation in Computer Training and Use in Secondary Schools: Interim Report*, 12–13.

52. Ibid., 7. Jean H. Danver and John M. Nevison, "Secondary School Use of the Time-Shared Computer at Dartmouth College," in *Proceedings of the May 14–16, 1969, Spring Joint Computer Conference*, AFIPS '69 (Spring) (Boston: ACM, 1969), 681–689, doi:10.1145/1476793.1476907.

53. Kurtz, *Demonstration and Experimentation in Computer Training and Use in Secondary Schools: Interim Report*, appendix A, Individual Student's Programs.

54. Kemeny and Kurtz, *The Dartmouth Time-Sharing Computing System*, 8.

55. John M. Nevison, *The Computer as Pupil: The Dartmouth Secondary School Project, Final Report*, October 1970, 16, Dartmouth College History Collection, Rauner Library.

56. Student version of *Spacewar!* in *NSF-Dartmouth Secondary School Project Monthly Bulletin* 3, no. 2 (1969): 7–8, reprinted in Nevison, *The Computer as Pupil*, appendix C. For more on *Spacewar!*, see Stewart Brand, "SPACEWAR: Fanatic Life and Symbolic Death among the Computer Bums," *Rolling Stone*, December 7, 1972.

57. "LOVE Program," in *NSF-Dartmouth Secondary School Project Monthly Bulletin* 3, no. 2 (1969): 8–9, reprinted in Nevison, *The Computer as Pupil*, appendix C.

58. Nevison, *The Computer as Pupil*, 19.

59. Ibid.

60. Ibid., 18.

61. *Dartmouth Alumni Magazine*, November 1967 issue, article "The Computer in Secondary Education," about the Secondary School Project. Information about Secondary School Project schools, the NSF grant, and its purpose in Information File for the Computation Center, Rauner Library; and (for example) *Kiewit Comments* 1, no. 5 (June 19, 1967); *Kiewit Comments* 2, no. 6 (August 19, 1968); *Kiewit Comments* 2, no. 9 (November 22, 1968); *Kiewit Comments* 3, no. 2 (February 20, 1969); *Kiewit Comments* 3, no. 5 (June 1, 1969); *Kiewit Comments* 3, no. 6 (September 20, 1969); and *Kiewit Comments* 4, no. 6 (September 28, 1970).

62. "Loomis Catches Up with Computer Age," *Loomis Bulletin*, July 1968, Loomis Chaffee School Archives; G. Albert Higgins, "BASIC: A New Language at Mount Hermon," *The Bulletin of Northfield and Mount Hermon Schools* (Spring 1967), clipping in Folder DA 29 (7841) 5 of 6, labeled: Miscellaneous Time-Sharing, Rauner Library.

63. Clippings from *Valley News*, *Caledonia Record*, *Recorder-Gazette* (Greenfield, MA), *Gazette* (Haverhill, MA), *Eagle* (Claremont, NH), *Evening Eagle-Tribune* (Lawrence, MA), *Herald* (Portsmouth, NH), *Union* (Springfield, MA), *Monitor & New Hampshire Patriot* (Concord, NH), *Telegraph* (Nashua, NH), *Sunday Globe* (Boston, MA), *Connecticut Valley Times Reporter* (Bellows Falls, VT), *Hartford Courant*, *Mascoma Week* (Canaan, NH), in Folder 1964–67, Box 4242, RDCCS, DA-181, Rauner Library. The language in these clippings is nearly identical, suggesting a common author; that is why these multiple clippings are cited in the next three notes.

64. Ibid.

65. Ibid.

66. Ibid.

67. In some cases, the young women at sister schools received computing access. For example, the girls at Northfield occasionally computed at Mount Hermon, per Higgins, "BASIC: A New Language at Mount Hermon."

68. Kurtz, *Demonstration and Experimentation in Computer Training and Use in Secondary Schools: Interim Report;* Danver and Nevison, "Secondary School Use of the Time-Shared Computer at Dartmouth College."

69. Nevison, *The Computer as Pupil; NSF-Dartmouth College Secondary School Project Biweekly Bulletin* 2, no. 7 (March 1969), Secondary School Folder, Box 4247, RDCCS, DA-181, Rauner Library.

70. *Kiewit Comments* 1, no. 6 (August 9, 1967); *Kiewit Comments* 5, no. 1 (January 11, 1971).

71. Goldin, Katz, and Kuziemko, "The Homecoming of American College Women."

72. Kurtz, *Demonstration and Experimentation in Computer Training and Use in Secondary Schools: Interim Report.*

73. *Kiewit Comments* 1, no. 3 (March 22, 1967): 2.

74. Bennet Vance in *Kiewit Comments* 2, no. 2 (February 19, 1968): 4, and *Kiewit Comments* 2, no. 5 (May 24, 1968): 3. Vance is, as of April 2018, employed at Dartmouth College's Neukom Institute for Computational Science.

75. *Kiewit Comments* 2, no. 1 (January 15, 1968): 2.

76. Nevison, *The Computer as Pupil*, 8. This seems to be unique users, or a close approximation based on distinct user IDs. At the public schools, the teletypes were accessible for eight hours each day. Schools seemed to regulate access in fifteen- or twenty-minute time slots, providing about twenty-four slots per day. Estimating twenty school days per month, the public schools had 480 slots for usage per month, so eighty unique users seems plausible.

77. Danver and Nevison, "Secondary School Use of the Time-Shared Computer at Dartmouth College," 686.

78. Kurtz, *Demonstration and Experimentation in Computer Training and Use in Secondary Schools: Interim Report.*

79. Ibid.; Danver, *Demonstration and Experimentation in Computer Training and Use in Secondary Schools, Interim Report II;* Nevison, *The Computer as Pupil.*

80. For example, one bulletin mentioned that "in Phase II BASIC, the symbols =<, =>, >< will be accepted as well as <=, >=, <>." In *NSF-Dartmouth College Secondary School Project Biweekly Bulletin* 2, no. 7 (March 1969): 1, Secondary School Folder, Box 4247, RDCCS, Rauner Library.

81. Individual winners received a "small silver abacus charm," and the school with the most points in two months received a silver cup. Kurtz, *Demonstration and Experimentation in Computer Training and Use in Secondary Schools: Interim Report*, 5.

82. Kurtz, *Demonstration and Experimentation in Computer Training and Use in Secondary Schools: Interim Report;* update on Kiewit contact person in *NSF–Dartmouth College Secondary School Project Biweekly Bulletin* 2, no. 7 (March 1969): 1, Secondary School Folder, Box 4247, RDCCS, Rauner Library.

83. Nevison, *The Computer as Pupil.*

84. Ibid., 28–29.

85. Ibid., 28.

86. Kemeny and Kurtz, *The Dartmouth Time-Sharing Computing System.*

87. Kurtz used "missionary" in the *Dartmouth College Regional College Consortium Final Report to the National Science Foundation 1970*, 7, Box 7977, RDCCS, Rauner Library. He later used "missionary" in the question-and-answer session at the 1978 History of Programming Languages Conference. Kurtz, "BASIC Session."

88. Kurtz, Rieser, and Meck, "Application to the National Science Foundation," 31–32.

89. Thomas E. Kurtz, *Interim Report of July 1969 on the Dartmouth College Regional College Consortium*, Box 7977, RDCCS, Rauner Library.

90. Kurtz, *Dartmouth College Regional College Consortium Final Report.*

91. Thomas E. Kurtz, *Report: Dartmouth College Establishment of a Model Regional Computer Center Using Time-Sharing, Grant GJ-11, March 68–June 69*, 3–4, Box 4, Garland Papers.

92. Ibid.

93. J. A. N. Lee, "The Rise and Fall of the General Electric Corporation Computer Department," *IEEE Annals of the History of Computing* 17, no. 4 (Winter 1995): 24–45.

94. David Ziegler, e-mail message to author, December 18, 2013.

95. Kurtz, *Dartmouth College Regional College Consortium Final Report*.

96. Dartmouth College Office of Information Services, April 6, 1971 Press Release, Folder 2 of 6, DA-29 (7841) DTSS Time-Sharing, Rauner Library.

97. Ibid.

98. Robert Albrecht, "A Modern-Day Medicine Show," *Datamation*, July 1963, http://www.rakahn.com/shared/medicine_show_1963.pdf. Details about Albrecht from Bob Johnstone, *Never Mind the Laptops: Kids, Computers, and the Transformation of Learning* (New York: iUniverse, 2003); Levy, *Hackers*; Swaine and Freiberger, *Fire in the Valley*.

99. Robert "Bob" Albrecht, e-mail message to the author, January 26, 2014.

100. National Council of Teachers of Mathematics, *Computer Facilities for Mathematics Instruction* (Washington, DC: National Council of Teachers of Mathematics, 1967); National Council of Teachers of Mathematics, *Introduction to an Algorithmic Language (BASIC)* (Washington, DC: National Council of Teachers of Mathematics, 1968).

101. National Council of Teachers of Mathematics, *Computer Facilities for Mathematics Instruction*, abstract.

102. Levy employs the language of evangelization in *Hackers: Heroes of the Computer Revolution*, such as Albrecht "spreading the gospel of the Hacker Ethic" (166), Albrecht as a "prophet of BASIC" (167), and "the mission of spreading computing to the people" (170).

103. Bob Albrecht, *My Computer Likes Me When I Speak in BASIC* (Menlo Park, CA: Dymax, 1972).

104. Ibid., 1.

105. Paul Sabin, *The Bet: Paul Ehrlich, Julian Simon, and Our Gamble over Earth's Future* (New Haven, CT: Yale University Press, 2013).

106. Donella H. Meadows, Dennis L. Meadows, Jørgen Randers, and William W. Behrens, *The Limits to Growth: A Report for the Club of Rome's Project on the Predicament of Mankind*, electronic ed. (New York: Universe Books, 1972), http://www.dartmouth.edu/~library/digital/publishing/meadows/ltg/; Jørgen Stig Nørgård, John Peet, and Kristín Vala Ragnarsdóttir, "The History of *The Limits to Growth*," http://donellameadows.org/archives/the-history-of-the-limits-to-growth/.

107. Fred Turner, *From Counterculture to Cyberculture: Stewart Brand, the Whole Earth Network, and the Rise of Digital Utopianism* (Chicago: University of Chicago Press, 2006).

108. Albrecht, *My Computer Likes Me When I Speak in BASIC*, 12.

109. *People's Computer Company* 1, no. 1 (October 1972): 1.

110. Ibid., back cover.

111. BAGELS in *People's Computer Company* 1, no. 1 (October 1972) and *People's Computer Company* 1, no. 2 (December 1972); MUGWUMP and HURKLE in *People's Computer Company* 1, no. 4 (April 1973); INCHWORM in *People's Computer Company* 1, no. 5 (May 1973).

112. Rick Beban, "Dear People" letter, *People's Computer Company* 2, no. 1 (September 1973): 2.

113. "People's Computer Center Info," *People's Computer Company* 2, no. 1 (September 1973): 3.

114. See, for example, *People's Computer Company* 2, no. 4 (March 1974): 22–23.

115. Music in (for example) *People's Computer Company* 1, no. 2 (December 1972); *People's Computer Company* 1, no. 3 (February 1973); and *People's Computer Company* 1, no. 5 (May 1973); "The Computer and the Artist" was *People's Computer Company* 2, no. 6 (July 1974).

116. *People's Computer Company* promoted the Huntington Project in issues 1, no. 1 (October 1972) and 1, no. 2 (December 1972). *People's Computer Company* also regularly provided information on DEC's EduSystem minicomputers (its line for the educational market) and on HP's minicomputers, as well as how to write bid specifications for such systems.

117. Martin Campbell-Kelly, William Aspray, Nathan Ensmenger, and Jeffrey R. Yost, *Computer: A History of the Information Machine*, 3rd ed. (Boulder, CO: Westview Press, 2014); Edgar H. Schein, *DEC Is Dead, Long Live DEC: The Lasting Legacy of Digital Equipment Corporation* (San Francisco: Berrett-Koehler, 2003).

118. Global estimate from "Internet History," Computer History Museum, http://www.computerhistory.org/internet_history/; DEC numbers from Campbell-Kelly et al., *Computer*, 217–218.

119. Johnstone, *Never Mind the Laptops*, 61.

120. Campbell-Kelly et al., *Computer;* Paul E. Ceruzzi, *A History of Modern Computing*, 2nd ed. (Cambridge, MA: MIT Press, 2003).

121. Ceruzzi, *A History of Modern Computing*, 191.

122. Kemeny and Kurtz, *BASIC Programming*.

123. *People's Computer Company* 1, no. 1 (October 1972): 15.

124. *People's Computer Company* 1, no. 2 (December 1972): 3.

125. *People's Computer Company* 1, no. 1 (October 1972): 14.

126. For DEC, see Ceruzzi, *A History of Modern Computing*, 129–136.

127. Marian Visich Jr. and Ludwig Braun, "The Use of Computer Simulations in High School Curricula," January 1974, http://eric.ed.gov/?id=ED089740.

128. "Computer Project Programs LI Highs," *Polytechnic Reporter* 61, no. 2 (September 26, 1968): 4, Bern Dibner Library, New York University Tandon School of Engineering.

129. Johnstone, *Never Mind the Laptops*, 56–60.

130. Leonard Rieser to John Dickey, July 18, 1966, Folder Phase II Early Papers, Box 4248, RDCCS, Rauner Library.

131. Braun quoted in Johnstone, *Never Mind the Laptops*, 57.

132. Visich and Braun, "The Use of Computer Simulations in High School Curricula."

133. Ibid.

134. Ludwig Braun and James Friedland, "Huntington II Simulation Program—BUFLO. Student Workbook, Teacher's Guide, and Resource Handbook," March 1974, http://eric.ed.gov/?id=ED179411.

135. Ludwig Braun, "BASIC Is Alive and—Well?," *SIGCUE Outlook* 5, no. 5 (October 1971): 217–220, doi:10.1145/965880.965882; Ludwig Braun, "Acceptance of Computers in Education," *SIGCUE Outlook* 5, no. 4 (August 1971): 171–174, doi:10.1145/965872.965876.

136. Beverly Hunter, *Learning Alternatives in U.S. Education: Where Student and Computer Meet* (Englewood Cliffs, NJ: Educational Technology Publications, 1975).

137. Theodor H. Nelson, *Computer Lib: You Can and Must Understand Computers Now/Dream Machines: New Freedoms through Computer Screens—a Minority Report* (Chicago: Nelson, 1974), 16–17; emphasis in original.

4 The Promise of Computing Utilities and the Proliferation of Networks

1. The so-called demise of computing utilities in the late 1960s figures most prominently in Martin Campbell-Kelly, William Aspray, Nathan Ensmenger, and Jeffrey R. Yost, *Computer: A History of the Information Machine*, 3rd ed. (Boulder, CO: Westview Press, 2014), 212.

2. John McCarthy, "Time-Sharing Computer Systems," in *Computers and the World of the Future*, ed. Martin Greenberger (Cambridge, MA: MIT Press, 1962), 221–236.

3. Sources for Whirlwind and SAGE include Kent C. Redmond, *From Whirlwind to MITRE: The R&D Story of the SAGE Air Defense Computer* (Cambridge, MA: MIT Press, 2000); Judy Elizabeth O'Neill, "The Evolution of Interactive Computing through Time-Sharing and Networking" (PhD diss., University of Minnesota, 1992); Arthur L. Norberg, Judy E. O'Neill, and Kerry J. Freedman, *Transforming Computer Technology: Information Processing for the Pentagon, 1962–1986* (Baltimore: Johns Hopkins University Press, 1996); Campbell-Kelly et al., *Computer;* Paul Edwards, *The Closed World: Computers and the Politics of Discourse in Cold War America* (Cambridge, MA: MIT Press, 1996); M. Mitchell Waldrop, *The Dream Machine: J. C. R. Licklider and the Revolution That Made Computing Personal* (New York: Penguin Books, 2002).

4. McCarthy, "Time-Sharing Computer Systems," 222.

5. Ibid., 229.

6. Estimate of six thousand computers from Charles Eames and Ray Eames, *A Computer Perspective*, ed. Glen Fleck (Cambridge, MA: Harvard University Press, 1973).

7. McCarthy, "Time-Sharing Computer Systems," 231.

8. Ibid., 236.

9. Ibid.

10. Ibid.

11. MIT's CTSS and Project MAC have been abundantly documented in both primary and secondary sources, such as Robert M. Fano, "The MAC System: The Computer Utility Approach," *IEEE Spectrum* 2, no. 1 (1965): 56–64; Fernando J. Corbató, Jerome H. Saltzer, and Chris T. Clingen, "Multics: The First Seven Years," in *Proceedings of the May 16–18, 1972, Spring Joint Computer Conference* (New York: ACM, 1972), 571–583; J. A. N. Lee, Robert Mario Fano,

Allan L. Scherr, Fernando J. Corbató, and Victor A. Vyssotsky, "Project MAC (Time-Sharing Computing Project)," *IEEE Annals of the History of Computing* 14, no. 2 (1992): 9–13; F. J. Corbató, oral history interview with Fernando J. Corbató (1990), Charles Babbage Institute, retrieved from the University of Minnesota Digital Conservancy, http://hdl.handle.net/11299/107230; Robert M. Fano, oral history interview with Robert M. Fano (1989), Charles Babbage Institute, retrieved from the University of Minnesota Digital Conservancy, http://hdl.handle.net/11299/107281; David Walden and Tom Van Vleck, eds., *The Compatible Time Sharing System (1961–1973)*, 50th Anniversary Commemorative Overview (Washington, DC: IEEE Computer Society, 2011), http://multicians.org/thvv/compatible-time-sharing-system.pdf; Steven Levy, *Hackers: Heroes of the Computer Revolution* (Sebastopol, CA: O'Reilly Media, 2010).

12. John L. Rudolph, *Scientists in the Classroom: The Cold War Reconstruction of American Science Education* (New York: Palgrave, 2002); Audra J. Wolfe, "Speaking for Nature and Nation: Biologists as Public Intellectuals in Cold War Culture" (PhD diss., University of Pennsylvania, 2002); Audra J. Wolfe, *Competing with the Soviets: Science, Technology, and the State in Cold War America* (Baltimore: Johns Hopkins University Press, 2013); Jamie Cohen-Cole, *The Open Mind: Cold War Politics and the Sciences of Human Nature* (Chicago: University of Chicago Press, 2016); Christopher J. Phillips, *The New Math: A Political History* (Chicago: University of Chicago Press, 2015).

13. J. C. R. Licklider, "Man–Computer Symbiosis," *IRE Transactions on Human Factors in Electronics* HFE-1 (1960): 4–11, https://groups.csail.mit.edu/medg/people/psz/Licklider.html.

14. Robert M. Fano and Fernando J. Corbató, "Time-Sharing on Computers," *Scientific American* 215, no. 3 (September 1966): 129–140.

15. Martin Greenberger, "The Computers of Tomorrow," *Atlantic Monthly*, May 1964, http://www.theatlantic.com/past/docs/unbound/flashbks/computer/greenbf.htm.

16. Bush, quoted in ibid.

17. Greenberger, "The Computers of Tomorrow."

18. Ibid.

19. Ibid.

20. Fano and Corbató, "Time-Sharing on Computers."

21. Ibid., 129.

22. Ibid.

23. Ibid., 140.

24. Ibid.

25. John McCarthy, "Information," *Scientific American* 215, no. 3 (September 1966): 65, 71.

26. Ibid., 71.

27. Campbell-Kelly et al., *Computer;* Christopher McDonald, "Building the Information Society: A History of Computing as a Mass Medium" (PhD diss., Princeton University, 2011).

28. Jeremy Main quoted in Campbell-Kelly et al., *Computer*, 210.

29. Dartmouth initially implemented its time-sharing system on a GE-225 computer in May 1964, which was upgraded to a GE-235 computer in

September 1964 per Dartmouth's agreement with GE. Detailed in [Thomas Kurtz], "Course Content Improvement Project Informal Report of 7 October 1964," Secondary School Project Folder, Box 4247, Records of Dartmouth College Computing Services (hereafter RDCCS), DA-181, Dartmouth College Rauner Special Collections Library (hereafter Rauner Library).

30. J. A. N. Lee, "The Rise and Fall of the General Electric Corporation Computer Department," *IEEE Annals of the History of Computing* 17, no. 4 (Winter 1995): 24–45, doi:10.1109/85.477434; Homer Oldfield, *King of the Seven Dwarfs: General Electric's Ambiguous Challenge to the Computer Industry* (Los Alamitos, CA: IEEE Computer Society Press, 1996).

31. John G. Kemeny, "The GE–Dartmouth Computer Partnership," DTSS History Birth 1967–70 Folder 2, Box 4, Stephen Garland Papers, ML-101, Rauner Library (hereafter Garland Papers). Corroborated by Anthony Knapp, e-mail messages to author, July 18 and 21, 2013.

32. Oldfield, *King of the Seven Dwarfs*, 143.

33. Ibid., 165.

34. [Kurtz], "Course Content Improvement Project Informal Report of 7 October 1964."

35. Ibid.; [Thomas Kurtz], "Progress Report Course Content Improvement Project of 15 December 1964," Secondary School Project Folder, Box 4247, RDCCS, Rauner Library.

36. [Kurtz], "Course Content Improvement Project Informal Report of 7 October 1964."

37. Ibid.; [Kurtz], "Progress Report Course Content Improvement Project of 15 December 1964"; Oldfield, *King of the Seven Dwarfs*, 191.

38. [Kurtz], "Progress Report Course Content Improvement Project of 15 December 1964."

39. Ibid.

40. Ibid.

41. [Kurtz], "Progress Report Course Content Improvement Project of 15 December 1964"; Transcript of the 1974 National Computer Conference Pioneer Day Session, Box 1, Papers of Sidney Marshall, MS-1054, Rauner Library; Kemeny, "The GE–Dartmouth Computer Partnership," Garland Papers.

42. Lee, "The Rise and Fall of the General Electric Corporation Computer Department," 36–37; Oldfield, *King of the Seven Dwarfs*, 191–192.

43. Lee, "The Rise and Fall of the General Electric Corporation Computer Department," 37.

44. Lee, "The Rise and Fall of the General Electric Corporation Computer Department"; Oldfield, *King of the Seven Dwarfs*.

45. *Now YOU Can Program a Computer in BASiC: A New Dimension in Data Processing*, General Electric Company brochure, 1965, Computer History Museum, http://www.computerhistory.org/brochures/doc-4372956d667f6/.

46. GE-*400 Time-Sharing Information Systems*, General Electric Company brochure, 1968, Computer History Museum, http://www.computerhistory.org/brochures/doc-4372956ed6cd1/.

47. Ibid.; emphasis added.

48. *Now YOU Can Program a Computer in BASiC.*

49. General Electric Information Systems Group, "Computer Time-Sharing on Campus: New Learning Power for Students," May 1968, Folder 3 of 6, DA-29 (7841) DTSS Time-Sharing, Rauner Library.

50. Ray Connolly, "GE's '67 Time-Sharing Gains," *Electronic News, Computer Trends*, Spring 1968, http://archive.computerhistory.org/resources/text/GEIS/102658913.1968.pdf.

51. General Electric Information Systems Group, "Computer Time-Sharing on Campus."

52. Lee, "The Rise and Fall of the General Electric Corporation Computer Department," 41.

53. Alex Conn, e-mail message to the author, December 3, 2013.

54. Ibid.

55. A. P. Conn, "Programming in BASIC Language," August 1968, Paris, France. Scanned manuscript e-mailed from Conn to the author.

56. Marc Bourin and Alex Conn, "Aspects de l'enseignement de l'informatique aux États-Unis," *Arts et Manufactures* 190 (October 1968): 43–46. Scanned article e-mailed from Conn to the author.

57. Kemeny and Kurtz believed, perhaps naively, that GE simply wanted to advance the time-sharing "state of the art," and would provide a computer in exchange for knowledge about the new time-sharing system. GE, of course, wanted an operable improved time-sharing system as quickly as possible for business purposes, with less regard for Kemeny and Kurtz's concerns about various computing constituencies. Compounding the tension of these two perspectives was that both Dartmouth and GE programmers underestimated the difficulty of developing a more powerful time-sharing system to serve more simultaneous users. See Kemeny, "The GE–Dartmouth Computer Partnership"; contents of Phase II Early Papers Folder, Box 4248, RDCCS, Rauner Library; Lee, "The Rise and Fall of the General Electric Corporation Computer Department"; Oldfield, *King of the Seven Dwarfs*.

58. Oldfield, *King of the Seven Dwarfs*, 172; "Kiewit Computation Center Fiscal 1967–68 Annual Report to the Faculty," Information File for the Computation Center, Rauner Library.

59. Warner R. Sinback, "Status Report, Dartmouth Project, 2 June 1966," Phase II Early Papers Folder, Box 4248, RDCCS, Rauner Library.

60. 1968 General Electric full-page color advertisement, Information File for the Computation Center, Rauner Library.

61. Ibid.

62. General Electric Information Systems Group, "Computer Time-Sharing on Campus."

63. United States President's Science Advisory Committee, Panel on Computers in Education (John R. Pierce et al.), *Computers in Higher Education: Report of the President's Science Advisory Committee* (Washington, DC: The White House, 1967).

64. General Electric Information Systems Group, "Computer Time-Sharing on Campus," 1.

65. Ibid., 8.

66. Campbell-Kelly et al., *Computer*, 212.

67. Lee, "The Rise and Fall of the General Electric Corporation Computer Department," 37; Martin Campbell-Kelly and Daniel D. Garcia-Swartz, *From Mainframes to Smartphones: A History of the International Computer Industry* (Cambridge, MA: Harvard University Press, 2015).

68. Lee, "The Rise and Fall of the General Electric Corporation Computer Department," 44.

69. Theodor H. Nelson, *Computer Lib: You Can and Must Understand Computers Now/Dream Machines: New Freedoms through Computer Screens—a Minority Report* (Chicago: Nelson, 1974).

70. John Kopf, "TYMNET as a Multiplexed Packet Network," in *Proceedings of the June 13–16, 1977 National Computer Conference* (New York: ACM, 1977), 609–613; Janet Abbate, "Government, Business, and the Making of the Internet," *Business History Review* 75, no. 1 (2001): 147–176.

71. Martin Campbell-Kelly and Daniel D. Garcia-Swartz, "Economic Perspectives on the History of the Computer Time-Sharing Industry, 1965–1985," *IEEE Annals of the History of Computing* 30, no. 1 (January 2008): 16–36.

72. Stuart H. Loory, "Computer Hookup to Home Foreseen," *New York Times,* December 5, 1966. Loory is reporting on Kemeny's speech at the Kiewit Center dedication "Conference on the Future Impact of Computers."

73. Ibid.

74. Greenberger, "The Computers of Tomorrow."

75. Ibid.

76. Ibid.

77. Ruth Schwartz Cowan, *More Work for Mother: The Ironies of Household Technology from the Open Hearth to the Microwave* (New York: Basic Books, 1983).

78. The scholarship on postwar American gender, sexuality, and family is extensive. Some key examples are Elaine Tyler May, *Homeward Bound: American Families in the Cold War Era,* fully revised and updated 20th anniversary edition (New York: Basic Books, 2008); Joanne Meyerowitz, ed., *Not June Cleaver: Women and Gender in Postwar America, 1945–1960* (Philadelphia: Temple University Press, 1994); Roger Horowitz and Arwen Mohun, eds., *His and Hers: Gender, Consumption and Technology* (Charlottesville: University Press of Virginia, 1998); Joanne Meyerowitz, "Beyond the Feminine Mystique: A Reassessment of Postwar Mass Culture, 1946–1958," *Journal of American History* 79, no. 4 (1993): 1455–1482.

79. Douglas Parkhill, *The Challenge of the Computer Utility* (Reading, MA: Addison-Wesley, 1966), 156.

80. Ibid., 160–161.

81. Ibid., 163–164.

82. Ibid., 159.

83. Paul Baran, "The Future Computer Utility," in *The Computer Impact,* ed. Irene Taviss (Englewood Cliffs, NJ: Prentice-Hall, 1970), 83.

84. Ibid., 84, 81.

85. Ibid., 84.

86. Ibid., 92.

87. Quote from Parkhill, *The Challenge of the Computer Utility,* 167. More on the National Data Center from Carl Kaysen, "The Privacy Question," in Taviss, *The Computer Impact,* 161–168.

88. These issues are all discussed in various essays in Martin Greenberger, ed., *Computers, Communications, and the Public Interest* (Baltimore: Johns Hopkins Press, 1971). See especially "Civil Liberties and Computerized Data Systems," "Property Rights under the New Technology," and "Developing National Policy for Computers and Communications."

89. Parkhill, *The Challenge of the Computer Utility*, 167.

90. Ibid.

91. Ibid., 167–168.

92. Ibid., 168.

93. Ibid.

94. Ibid., 181.

95. Greenberger, *Computers, Communications, and the Public Interest*, vii.

96. Kemeny's "Large Time-Sharing Networks" presentation was likely on or around September 11, 1969, because a manuscript version of it with that date appears in the DTSS Info Publications folder, Box 4, Garland Papers.

97. Greenberger, *Computers, Communications, and the Public Interest*, xv.

98. John Kemeny, "Large Time-Sharing Networks," in Greenberger, *Computers, Communication, and the Public Interest*, 14.

99. Ibid., 15.

100. Ibid.

101. Ibid., 18, 22.

102. Ibid., 8.

103. Ibid., 8–9.

104. Kemeny later elaborated on the shoe-shopping example in his book *Man and the Computer* (New York: Scribner, 1972). He explained that his sister wore "rather narrow shoes" (like a 7½ AA) and that she often visited a dozen stores to find "a single pair in her size." He continued, "She would be grateful if before she left her home she could find out which stores have any shoes that might fit her" (122). Now, this is certainly not the image of an avid shoe-shopper, but Kemeny did not include this level of detail in his Hopkins-Brookings lecture. Rather, his invocation of a housewife looking for shoes invoked a more general American cultural trope.

105. Kemeny, "Large Time-Sharing Networks," 11.

106. Ibid.

107. Ibid., 6, 13.

108. Ibid., 9.

109. Greenberger, *Computers, Communications, and the Public Interest*, xviii; emphasis in original.

110. Carle Hodge, ed., "EDUCOM Means Communication," *EDUCOM: Bulletin of the Interuniversity Communications Council* 1, no. 1 (January 1966): inside cover.

111. Ibid.

112. "Remarkably swift progression . . ." in H. Eugene Kessler, "Education, Computing and Networking: A Review and Prospect," *EDUCOM: Bulletin of the Interuniversity Communications Council* 9, no. 1 (Spring 1974): 2; "to help identify . . ." in Martin Greenberger, Julius Aronofsky, James L. McKenney, and William F. Massy, eds., *Networks for Research and Education: Sharing Computing and Information Resources Nationwide* (Cambridge, MA: MIT Press, 1974), xii.

113. EDUCOM was not alone in identifying the swift growth of networks—and their importance—during this time. A sample of representative publications include Norman R. Nielsen, "The Merit of Regional Computing Networks," *Communications of the ACM* 14, no. 5 (May 1971); Andrew R. Molnar, "The Computer and the Fourth Revolution" (paper presented at the Association for Educational Data Systems Annual Convention, New Orleans, LA, April 16–19, 1973), http://files.eric.ed.gov/fulltext/ED087417.pdf; Martin Greenberger, Julius Aronofsky, James L. McKenney, and William F. Massy, "Computer and Information Networks," *Science* 182, no. 4107 (October 5, 1973): 29–35; Arthur W. Luehrmann and John M. Nevison, "Computer Use under a Free-Access Policy," *Science* 184, no. 4140 (May 31, 1974): 957–961, http://www.jstor.org/stable/1738321; Ronald W. Cornew and Philip M. Morse, "Distributive Computer Networking: Making It Work on a Regional Basis," *Science* 189, no. 4202 (1975): 523–531, http://www.jstor.org/stable/1740965.

114. Kessler, "Education, Computing and Networking," 2.

115. Carle Hodge, ed., "Networks for Instant Information," *EDUCOM: Bulletin of the Interuniversity Communications Council* 1, no. 7 (October 1966): 8.

116. Kessler, "Education, Computing and Networking," 2.

117. Greenberger et al., *Networks for Research and Education*, 5.

118. Norman Abramson, "Digital Broadcasting in Hawaii—The ALOHA System," *EDUCOM: Bulletin of the Interuniversity Communications Council* 9, no. 1 (Spring 1974): 9–13.

119. Louis T. Parker Jr. and Joseph R. Denk, "A Network Model for Delivering Computer Power and Curriculum Enhancement for Higher Education: The North Carolina Educational Computer Service," *EDUCOM: Bulletin of the Interuniversity Communications Council* 9, no. 1 (Spring 1974): 24–30.

120. Gerard P. Weeg and Charles R. Shomper, "The Iowa Regional Computer Network," *EDUCOM: Bulletin of the Interuniversity Communications Council* 9, no. 1 (Spring 1974): 14–15.

121. Thomas E. Kurtz, *Report: Dartmouth College Establishment of a Model Regional Computer Center Using Time-Sharing, Grant GJ-11, March 68–June 69*, Box 4, Garland Papers.

5 How *The Oregon Trail* Began in Minnesota

1. "*The Oregon Trail*, Inducted 2016," The Strong, National Museum of Play, World Video Game Hall of Fame, http://www.worldvideogamehalloffame.org/games/oregon-trail.

2. See, for example, Anna Garvey, "The Oregon Trail Generation: Life before and after Mainstream Tech," *Social Media Week*, April 21, 2015, https://socialmediaweek.org/blog/2015/04/oregon-trail-generation/; Steve Pepple, "Why Did All Children of a Certain Age Play Oregon Trail?," *Medium*, November 30, 2016, https://medium.com/@stevepepple/why-did-all-children-of-a-certain-age-play-oregon-trail-6a53e27e83d8. The author is one such individual whose earliest memory of computing is playing *The Oregon Trail* on a Commodore 64 in elementary school.

3. I started researching this book in April 2011, and in the preceding months, two such articles had come to my attention: Jessica Lussenhop, "Oregon Trail: How Three Minnesotans Forged Its Path," *City Pages*, January 19, 2011, http://www.citypages.com/news/oregon-trail-how-three-minnesotans-forged-its-path-6745749; Josh Rothman, "The Origins of 'The Oregon Trail,'" *Boston Globe*, March 21, 2011, http://archive.boston.com/bostonglobe/ideas/brainiac/2011/03/the_origins_of_2.html. Since then, others have appeared with greater frequency than I would have expected: Dennis Scimeca, "Why *Oregon Trail* Still Matters," *The Kernel*, week of August 17, 2014, http://kernelmag.dailydot.com/issue-sections/features-issue-sections/10021/the-lasting-legacy-of-oregon-trail/; "The Story of *Oregon Trail*," *Weird H!story Podcast*, March 10, 2016, http://www.interestingtimespodcast.com/2016/71-live-at-the-jack-london-the-story-of-oregon-trail/; Matt Jancer, "How You Wound Up Playing *The Oregon Trail* in Computer Class," *Smithsonian Magazine Online*, July 22, 2016, http://www.smithsonianmag.com/innovation/how-you-wound-playing-em-oregon-trailem-computer-class-180959851/; Kevin Wong, "The Forgotten History of 'The Oregon Trail,' as Told by Its Creators," *Motherboard*, February 15, 2017, https://motherboard.vice.com/en_us/article/qkx8vw/the-forgotten-history-of-the-oregon-trail-as-told-by-its-creators; Jaz Rignall, "A Pioneering Game's Journey: The History of *Oregon Trail*," *US Gamer*, April 19, 2017, http://www.usgamer.net/articles/the-oral-history-of-oregon-trail.

4. Minnesota School Districts Data Processing Joint Board (widely known and hereafter cited as TIES), *TIES & TALES* 1, no. 1 (September 15, 1967), Folder for *TIES & TALES*, Box 1, Total Information for Educational Systems (TIES) Records, Charles Babbage Institute. All issues of the *TIES & TALES* newsletters cited in this book were drawn from this *TIES & TALES* folder (in Box 1, TIES Records, Charles Babbage Institute), so archive information is omitted from subsequent citations.

5. John E. Haugo, "Minnesota Educational Computing Consortium," April 1973, http://eric.ed.gov/?id=ED087434.

6. Minnesota Educational Computing Consortium (widely known and hereafter cited as MECC), *Systems Update* 2, no. 1 (February 1975): 3, microfilm, Minnesota Historical Society.

7. Dale Eugene LaFrenz, "The Documentation and Formative Evaluation of a Statewide Instructional Computing Network in the State of Minnesota" (PhD diss., University of Minnesota, 1978), 367.

8. Nathan Ensmenger, "Power to the People: Toward a Social History of Computing," *IEEE Annals of the History of Computing* 26, no. 1 (2004): 96, 94–95; Joy Rankin, "Toward a History of Social Computing: Children, Classrooms, Campuses, and Communities," *IEEE Annals of the History of Computing* 36, no. 2 (2014): 88, 86–87.

9. Thomas J. Misa, *Digital State: The Story of Minnesota's Computing Industry* (Minneapolis: University of Minnesota Press, 2013).

10. Thomas J. Misa, "Digital State: The Emergence of Minnesota's High-Technology Economy" (presented at the Massachusetts Institute of Technology HASTS Colloquium, Cambridge, MA, October 22, 2012).

11. Donald Bitzer, oral history interview by Sheldon Hochheiser, February 19, 1988, Charles Babbage Institute, retrieved from the University of Minnesota Digital Conservancy, http://hdl.handle.net/11299/107121.

12. Self-description in Dale Eugene LaFrenz, oral history interview by Judy E. O'Neill, April 13, 1995, Charles Babbage Institute, retrieved from the University of Minnesota Digital Conservancy, http://hdl.handle.net/11299/107423. My evaluation of his salesmanship based on Dale Eugene LaFrenz, conversation with the author, May 28, 2013, Minneapolis, MN.

13. Bonnee Fleming et al., eds., *Bisbila 1963–64* [yearbook], 12–13, Box 4, University High School Records, University of Minnesota Archives (hereafter University High Records).

14. *University High School: A Report on the Purposes, Programs, Projects and Pupils of a Campus Laboratory School* (Minneapolis, MN, 1967), 3, Information File for University High School, University of Minnesota Archives.

15. Ibid., 6.

16. David C. Johnson, Larry L. Hatfield, Dale E. LaFrenz, and Thomas E. Kieren, *Preliminary Report on Computer Mediated Instruction in Mathematics*, February 1, 1966, Folder for Courses of Study, Box 1, University High Records; LaFrenz, interview.

17. LaFrenz, interview, 7.

18. Johnson et al., *Preliminary Report on Computer Mediated Instruction in Mathematics*.

19. Ibid.; David C. Johnson, Larry L. Hatfield, Dale E. LaFrenz, Thomas E. Kieren, John W. Walther, and Kenneth W. Kelsey, *Preliminary Report #2 on Computer Assisted Mathematics Program, CAMP*, December 1, 1966, Folder for Courses of Study, Box 1, University High Records.

20. Johnson et al., *Preliminary Report on Computer Mediated Instruction in Mathematics*, Grade Nine Section.

21. Ibid., Grades Seven and Eleven Sections.

22. Larry L. Hatfield, "Computer Mediated Mathematics: A Ripple or a Wave?" (reprinted from Minnesota Council of Teachers of Mathematics [MCTM] Newsletter, May 1966), Information File for University High School, University of Minnesota Archives; emphasis added.

23. Richard P. Kleeman, "University High Is 'Lab' for Curriculums That Are 'Far Out,'" *Minneapolis Tribune*, February 27, 1966, Information File for University High School, University of Minnesota Archives.

24. *University High School.*

25. Ibid., 3–4, 16.

26. *Bisbila 1966–67* [yearbook], 14–15, Box 4, University High Records.

27. Ibid., 12.

28. Mary Lee Wright, ed., *Bisbila 1967–68* [yearbook], Box 2, University High Records.

29. LaFrenz, interview, 9.

30. Becky Hall, "Computer Hook-up Opens Door for Math Teaching Experiment," *Campus Breeze* 48, no. 3 (December 17, 1965), Box 5, University High Records.

31. Jim Cohen, "Computer Implements Math Classes," *Campus Breeze* 49, no. 2 (November 18, 1966), Box 5, University High Records.

32. "A Glance Around," *Campus Breeze* 50, no. 3 (November 22, 1967), Box 5, University High Records.

33. "Computer Game Intrigues Students and Teachers," *Campus Breeze* 50, no. 7 (April 26, 1968), Box 5, University High Records.

34. Ibid.

35. Ibid.

36. Johnson et al., *Preliminary Report on Computer Mediated Instruction in Mathematics.*

37. Johnson et al., *Preliminary Report #2 on Computer Assisted Mathematics Program, CAMP,* 29.

38. "Pillsbury Co. Purchases GE Computer," *Schenectady Gazette,* May 1, 1965; "Pillsbury Forms Unit to Offer Wide Range of Computer Services," *Wall Street Journal,* January 16, 1968.

39. Johnson et al., *Preliminary Report #2 on Computer Assisted Mathematics Program, CAMP.*

40. LaFrenz, interview, 9.

41. Liz Burkhardt, ed., *Bisbila 1965–66* [yearbook], 15, Box 4, University High Records.

42. Reprinted in Johnson et al., *Preliminary Report on Computer Mediated Instruction in Mathematics.*

43. Hatfield, "Computer Mediated Mathematics."

44. "Math Teachers Visit 'Vegas,'" *Campus Breeze* 49, no. 7 (April 21, 1967), Box 5, University High Records.

45. Jo Freeman, "On the Origins of Social Movements," in *Waves of Protest: Social Movements since the Sixties,* ed. Jo Freeman and Victoria Johnson (Lanham, MD: Rowman & Littlefield, 1999), 7–24.

46. Characteristics of a social movement discussed in Jo Freeman and Victoria Johnson, eds., *Waves of Protest: Social Movements since the Sixties* (Lanham, MD: Rowman & Littlefield, 1999), introduction. Other helpful works on social movements include Terry H. Anderson, *The Movement and the Sixties* (New York: Oxford University Press, 1995); Charles Tilly, *Social Movements, 1768–2004* (Boulder, CO: Paradigm, 2004); Vincenzo Ruggiero and Nicola Montagna, eds., *Social Movements: A Reader* (London: Routledge, 2008); Jeff Goodwin and James M. Jasper, eds., *The Social Movements Reader: Cases and Concepts* (Malden, MA: Blackwell, 2003).

47. Educational Research and Development Council of the Twin Cities Metropolitan Area, *Prospectus TIES: Total Information for Educational Systems* (Minneapolis, MN: Educational Research and Development Council, 1967), Census Questionnaire Section.

48. Ibid., numbered page 2. Not all pages in this publication are numbered, so I have provided a page number when there is one, and a section when there is no page number.

49. Ibid., part V, section 4.

50. Ibid., numbered page 7.

51. The joint board recognized early on that its name was rather "cumbersome," and it resolved to adopt Total Information for Educational Systems, with the acronym TIES, as its title "for the purposes of identifying more easily with Joint Board constituency and the general public." From TIES Executive Committee Meeting Minutes for December 20, 1967, 5, Folder for TIES Committee Minutes, Box 1, Total Information for Educational Systems (TIES) Records, Charles Babbage Institute (hereafter cited as TIES Records).

52. Educational Research and Development Council of the Twin Cities Metropolitan Area, *Prospectus TIES*, introduction 1.

53. Ibid., numbered page 3.

54. Ibid., numbered page 4.

55. Ibid.

56. Ibid., numbered pages 9 and 33.

57. TIES, *TIES & TALES* 1, no. 2 (November 1967): 1.

58. Such meetings are mentioned, for example, in *TIES & TALES* 1, no. 5 (June 1968); *TIES & TALES* 1, no. 6 (September 1968); *TIES & TALES* 2, no. 1 (February 1969); *TIES & TALES* 2, no. 2 (July 1969); *TIES & TALES* 3, no. 1 (November 1969); *TIES & TALES* 3, no. 3 (June 1970).

59. TIES, *TIES & TALES* 1, no. 5 (June 1968).

60. TIES, *TIES & TALES* 4, no. 3 (February 1971).

61. TIES, *TIES & TALES* 1, no. 1 (September 1967).

62. TIES, *TIES & TALES* 4, no. 7 (May 1972).

63. TIES, *TIES & TALES* 1, no. 1 (September 1967): 2.

64. TIES, *TIES & TALES* 1, no. 4 (March 1968).

65. TIES, *TIES & TALES* 1, no. 6 (September 1968).

66. Ibid.

67. TIES, *TIES & TALES* 4, no. 5 (June 1971): 7.

68. TIES, *TIES & TALES* 3, no. 2 (April 1970): 5.

69. TIES, *TIES & TALES* 4, no. 3 (February 1971): 4.

70. TIES, *TIES & TALES* 4, no. 2 (December 1970).

71. TIES, *TIES & TALES* 4, no. 1 (October 1970): 9.

72. TIES, *TIES & TALES* 5, no 2. (May 1973): 15.

73. TIES (Donald Holznagel et al.), *Timely TIES Topics* 1, no. 1 (September 29, 1972), Folder for TIES Newsletter TTT, Box 1, TIES Records. All of the issues of the *Timely TIES Topics* newsletters cited in this book were drawn from this TIES Newsletter TTT folder (in Box 1, TIES Records, Charles Babbage Institute), so archive information is omitted from subsequent citations.

74. Ibid.

75. TIES, *TIES & TALES* 1, no. 6 (September 1968): 5.

76. TIES, *TIES & TALES* 3, no. 3 (June 1970): 1–2.

77. Ibid., 3–4.

78. Ibid., 4–5.

79. TIES, *TIES & TALES* 4, no. 1 (October 1970): 2.

80. Ibid., 1–2.

81. Ibid., 1.

82. Kristin Serum, "Computer System Installed in Schools," *Minneapolis Star*, January 19, 1967, clipping in Folder 1964–67, Box 4242, Records of Dart-

mouth College Computing Services, DA-181, Dartmouth College Rauner Special Collections Library (hereafter Rauner Library).

83. *Kiewit Comments* 1, no. 5 (June 19, 1967), Rauner Library.

84. LaFrenz, interview; Lussenhop, "Oregon Trail."

85. TIES, *TIES & TALES* 4, no. 6 (November 1971); Jean Danver, "Publication of Computer Curriculum Materials at Hewlett Packard Company," in *Computing and the Decision Makers: Where Does Computing Fit in Institutional Priorities?*, Proceedings of the EDUCOM Spring Conference, April 17–18, 1974 (Princeton, NJ: EDUCOM, 1974), 228–232.

86. Lussenhop, "Oregon Trail."

87. Donald Holznagel, Dale E. LaFrenz, and Michael Skow, "Historical Development of Minnesota's Instructional Computing Network," in *Proceedings of the 1975 Annual Conference* (New York: ACM, 1975), 79–80, doi:10.1145/800181.810282.

88. Ibid.; Kathryn L. Nelson, "An Evaluation of the Southern Minnesota School Computer Project, Presented in Foundations of Secondary Education 6353" (master's thesis, Mankato [Minnesota] State College, July 15, 1971).

89. James Armin Sydow, "Secondary Athletics and the Computer," January 1974, http://eric.ed.gov/?id=ED096971; James Armin Sydow, "Computers in Physical Education," November 1974, http://eric.ed.gov/?id=ED096972; Ross Taylor, "Learning with Computers: A Test Case," *Design Quarterly* 90/91 (January 1, 1974): 35–36.

90. Sydow, "Secondary Athletics and the Computer."

91. Haugo, "Minnesota Educational Computing Consortium."

92. Ibid.

93. Holznagel, LaFrenz, and Skow, "Historical Development of Minnesota's Instructional Computing Network."

94. Ibid.

95. W. G. Shepherd, "Response from the University of Minnesota Community to the Proposal for a Statewide Educational Computer Services Organization," October 16, 1972, 2, Folder MECC 1972–73, Box 58, Office of the Vice President for Academic Administration Records, University of Minnesota Archives.

96. Ibid.

97. TIES, *TIES & TALES* 5, no. 2 (May 1973): 14.

98. Alfred D. Hauer, "Major District Concerns Regarding MECC Proposal," October 12, 1972, Folder MECC 1972–73, Box 58, Office of the Vice President for Academic Administration Records, University of Minnesota Archives.

99. "January 22, 1973 Joint and Cooperative Agreement Minnesota Educational Computing Consortium," January 22, 1973, Folder MECC 1972–73, Box 58, Office of the Vice President for Academic Administration Records, University of Minnesota Archives.

100. Ibid.

101. Ibid.

102. Ibid.

103. Peter Roll, *A Brief Report on the Status of Planning for the Minnesota Educational Computer Consortium (MECC)*, February 6, 1973, Folder MECC 1972–73,

Box 58, Office of the Vice President for Academic Administration Records, University of Minnesota Archives.

104. Ad Hoc Committee for the MECC User's Association, "MECC User's Association Proposed Articles of Organization," 1975, Information File for Computer Center, University of Minnesota Archives.

105. TIES, *Timely TIES Topics* 3, no. 4 (November 1974); TIES, *Timely TIES Topics* 3, no. 6 (February 1975); TIES, *Timely TIES Topics* 3, no. 7 (March 1975).

106. Ad Hoc Committee for the MECC User's Association, "MECC User's Association Proposed Articles of Organization"; emphasis in original.

107. MECC, *Systems Update* 1, no. 2 (July 1974), microfilm, Minnesota Historical Society.

108. Ibid., 5.

109. Ibid.

110. MECC, *Systems Update* 1, no. 2 (July 1974); MECC, *Systems Update* 1, no. 3 (October 1974), microfilm, Minnesota Historical Society.

111. MECC, *Systems Update* 1, no. 2 (July 1974): 1.

112. Holznagel, LaFrenz, and Skow, "Historical Development of Minnesota's Instructional Computing Network."

113. Ibid.; MECC, *Systems Update* 1, no. 3 (October 1974); MECC, *Systems Update* 2, no. 1 (January–February 1975), microfilm, Minnesota Historical Society.

114. MECC, *Systems Update* 1, no. 2 (July 1974): 2.

115. Ibid.

116. MECC, *Systems Update* 1, no. 3 (October 1974): 1–2.

117. MECC, *Systems Update* 1, no. 4 (December 1974): 1, microfilm, Minnesota Historical Society.

118. MECC, *Systems Update* 1, no. 3 (October 1974): 1–2.

119. MECC, *Systems Update* 1, no. 2 (July 1974): 1.

120. MECC, *Systems Update* 1, no. 3 (October 1974): 1.

121. MECC, *Systems Update* 1, no. 2 (July 1974): 1–2.

122. MECC, *Systems Update* 2, no. 1 (January–February 1975): 3.

123. MECC, *Annual Report for 1977–78*, Folder for MECC 1975–80, Box 145, Office of the President Records, University of Minnesota Archives.

124. Ibid., 13.

125. Ibid.

126. TIES, *Timely TIES Topics* 2, no. 7 (March 1974): 4.

127. TIES, *Timely TIES Topics* 2, no. 6 (February 1974): 2.

128. Ibid., 3.

6 PLATO Builds a Plasma Screen

1. During the 1950s and 1960s, Mercy Hospital School of Nursing was located on West Park Street in Urbana, across the street from the hospital. From the article by Doris Hoskins and Sister Esther Matthew, "Mercy Hospital School of Nursing," *Through the Years: African American History in Champaign County* (Fall/Winter 1999): 1–2, http://www.museumofthegrandprairie.org/Images/TTY%20Fall%20-%20Winter%201999.pdf, archived at perma.cc/22EK-AKXS.

Bitzer reported on her nursing course in the PLATO Quarterly Progress Report for March–May 1963, Box 22, Collection CBI 133: University of Illinois at Urbana-Champaign, Computer-Based Education Research Laboratory PLATO Reports, PLATO Documents and CERL Progress Reports (hereafter CBI PLATO Collection), Charles Babbage Institute, Minneapolis, Minnesota; and in Maryann Bitzer, "Self-Directed Inquiry in Clinical Nursing Instruction by Means of the PLATO Simulated Laboratory," December 1963, Box 1, CBI PLATO Collection.

2. Donald Bitzer, Peter Braunfeld, and Wayne Lichtenberger, "PLATO: An Automatic Teaching Device," June 1961, CSL Report I-103, Box 1, CBI PLATO Collection.

3. PLATO Quarterly Progress Report for September 1966–February 1967, Box 22, CBI PLATO Collection; Elisabeth R. Lyman, "A Descriptive List of PLATO Programs: 1960–1968," CERL Publication X-2, May 1968, revised May 1970, Box 3, CBI PLATO Collection.

4. Elisabeth R. Lyman, "PLATO Highlights. Third Revision," December 1975, http://files.eric.ed.gov/fulltext/ED124143.pdf, archived at perma.cc/2ZG6-S7WY.

5. Donald L. Bitzer, "Signal Amplitude Limiting and Phase Quantization in Antenna Systems" (PhD diss., University of Illinois, 1960).

6. Coordinated Science Lab at the University of Illinois, "1950s: The Classified Years," http://csl.illinois.edu/about-lab/1950s-classified-years, archived at perma.cc/N3PC-5XGN.

7. Bethany Anderson, "The Birth of the Computer Age at Illinois," University of Illinois Archives, 2013, http://archives.library.illinois.edu/blog/birth-of-the-computer-age, archived at perma.cc/RJ57–9NV8.

8. Donald Bitzer, oral history interview by Mollie Price on August 17, 1982, Charles Babbage Institute.

9. Bitzer, interview; PLATO Quarterly Progress Report for June–August 1960, Box 22, CBI PLATO Collection.

10. Bitzer, interview, 7–8.

11. Donald Bitzer and Peter Braunfeld, "Automated Teaching Machine (PLATO)," Patent disclosure, April 1961, Box 1, CBI PLATO Collection.

12. PLATO Quarterly Progress Reports for June–August 1960 and September–November 1960, Box 22, CBI PLATO Collection; Donald Bitzer, Peter Braunfeld, and Wayne Lichtenberger, "PLATO: An Automatic Teaching Device," June 1961, CSL Report I-103, Box 1, CBI PLATO Collection. CSL Report I-103 was published as an article with the same title in *IRE Transactions on Education* E-4 (December 1961): 157–161.

13. Donald Bitzer, Peter Braunfeld, and Wayne Lichtenberger, "PLATO: An Automatic Teaching Device," *IRE Transactions on Education* E-4 (December 1961): 157–161.

14. This strategy of referencing existing media (books, blackboards) when introducing a new visual medium (PLATO), thereby creating continuity and familiarity, has a long history that persists to the present. For example, when I look at my iPhone, I see pictures of an analog clock, a 35mm camera, and a speech bubble to symbolize my Clock, Camera, and Messages applications. See

Jay David Bolter and Richard Grusin, *Remediation: Understanding New Media* (Cambridge, MA: MIT Press, 1998).

15. Bitzer, Braunfeld, and Lichtenberger, "PLATO: An Automatic Teaching Device" (December 1961), 158.

16. Ibid., 159. We now understand software as the programs that run on computers. During the 1960s, that concept of software did not yet exist; computer programs were inseparable from the machines on which they ran. See Martin Campbell-Kelly, *From Airline Reservations to Sonic the Hedgehog: A History of the Software Industry* (Cambridge, MA: MIT Press, 2003). However, I am using the term "software" here to describe the set of computer instructions and programs, and the digitized instructional materials that a student user encountered on the PLATO system. That is, I employ the term "software" to differentiate what happened on the TV display from the hardware of the display (and keyset) itself.

17. CSL Progress Report for December 1959–February 1960, Box 1, CBI PLATO Collection; Bitzer, Braunfeld, and Lichtenberger, "PLATO: An Automatic Teaching Device" (December 1961).

18. CSL Progress Report for March–May 1960 in Coordinated Science Laboratory, "Progress Report for 1959–1960," Illinois Digital Environment for Access to Learning and Scholarship, http://hdl.handle.net/2142/28161.

19. Bitzer, interview, 21–22.

20. Paul N. Edwards, *The Closed World: Computers and the Politics of Discourse in Cold War America* (Cambridge, MA: MIT Press, 1996); Stuart W. Leslie, *The Cold War and American Science: The Military-Industrial-Academic Complex at MIT and Stanford* (New York: Columbia University Press: 1993). Edwards and Leslie argue that in many cases, the Cold War dictated, and limited, the scientific research conducted at American universities.

21. PLATO Quarterly Progress Report for December 1960–February 1961, Box 22, CBI PLATO Collection.

22. University High School first discussed in PLATO Quarterly Progress Report for December 1960–February 1961, Box 22, CBI PLATO Collection; Committee on School Mathematics first discussed in PLATO Quarterly Progress Report for September–November 1962, Box 22, CBI PLATO Collection; Nursing project first discussed in PLATO Quarterly Progress Report for March–May 1963, Box 22, CBI PLATO Collection.

23. Conference presentations discussed in PLATO Quarterly Report for June–August 1961, Box 22, CBI PLATO Collection. Articles include Bitzer, Braunfeld, and Lichtenberger, "PLATO: An Automatic Teaching Device"; P. G. Braunfeld and L. D. Fosdick, "The Use of an Automatic Computer System in Teaching," *IRE Transactions on Education* 5, no. 3 (September 1962): 156–167.

24. Maryann Bitzer, "Self-Directed Inquiry in Clinical Nursing Instruction by Means of the PLATO Simulated Laboratory," 1963, Box 1, CBI PLATO Collection.

25. Ibid., 7.

26. Donald L. Bitzer, Elisabeth R. Lyman, and John A. Easley, "The Uses of PLATO: A Computer Controlled Teaching System," in *Proceedings of the 1966 Clinic on Library Applications of Data Processing*, ed. H. Goldhor (Urbana, IL:

Graduate School of Library Science, 1966), 34–46, Illinois Digital Environment for Access to Learning and Scholarship, http://hdl.handle.net/2142/1760.

27. Maryann Bitzer, "Clinical Nursing Instruction via the PLATO Simulated Laboratory," reprinted from *Nursing Research* 15, no. 2 (Spring 1966), Box 2, CBI PLATO Collection, no page numbers.

28. Ibid.

29. Bitzer, "Self-Directed Inquiry in Clinical Nursing Instruction by Means of the PLATO Simulated Laboratory," 15.

30. Ibid., 16.

31. PLATO Quarterly Progress Reports September–November 1960 and December 1960–February 1961, Box 22, CBI PLATO Collection.

32. Bitzer, interview, 10–11.

33. Ibid.

34. Bitzer and Braunfeld, "Automated Teaching Machine (PLATO)"; Bitzer, Braunfeld, and Lichtenberger, "PLATO: An Automatic Teaching Device"; PLATO Quarterly Progress Report for March–May 1961 (Box 22, CBI PLATO Collection) reported the addition of a "refresh" key, but Bitzer repeatedly referred to this as the RENEW key in his 1982 interview. By 1969 the key had been labeled "REPLOT," according to R. A. Avner and Paul Tenczar, *The TUTOR Manual*, 11 and 13, CERL Report X-4, January 1969, Box 3, CBI PLATO Collection.

35. Bitzer, interview, 14.

36. Ibid.

37. Ibid.

38. PLATO Quarterly Progress Reports for March–May 1961 and June–August 1961, Box 22, CBI PLATO Collection.

39. PLATO Quarterly Progress Report for June–August 1961, Box 22, CBI PLATO Collection.

40. PLATO Quarterly Progress Report for September–November 1962, Box 22, CBI PLATO Collection.

41. Fernando J. Corbató, Marjorie Merwin-Daggett, and Robert C. Daley, "An Experimental Time-Sharing System," in *Proceedings of the May 1–3, 1962, Spring Joint Computer Conference*, AIEE-IRE '62 (Spring) (New York: ACM, 1962), 335–344.

42. PLATO Quarterly Progress Report for September–November 1962, Box 22, CBI PLATO Collection.

43. PLATO Progress Report for March–August 1966, Box 22, CBI PLATO Collection.

44. PLATO Quarterly Progress Report for December 1962–February 1963, Box 22, CBI PLATO Collection, 30.

45. PLATO Quarterly Progress Report for September–November 1963, Box 22, CBI PLATO Collection, 60.

46. Ibid.

47. Edward Herman Stredde, "The Development of a Multicolor Plasma Display Panel," Folder 67, no. 01, Box 2, CBI PLATO Collection, 2.

48. PLATO Quarterly Progress Report for March–May 1964, Box 22, CBI PLATO Collection, 58.

49. PLATO Quarterly Progress Report for June–August 1964, Box 22, CBI PLATO Collection, 40.

50. PLATO Quarterly Progress Report for September–November 1964, Box 22, CBI PLATO Collection, 31.

51. Ibid.

52. Ibid., 40.

53. PLATO Quarterly Progress Report for June–August 1965, Box 22, CBI PLATO Collection.

54. PLATO Quarterly Progress Reports for September–November 1965 and December 1965–February 1966, Box 22, CBI PLATO Collection.

55. PLATO Progress Report for September 1966–February 1967, Box 22, CBI PLATO Collection.

56. B. M. Arora, D. L. Bitzer, H. G. Slottow, and R. H. Willson, "The Plasma Display Panel—A New Device for Information Display and Storage," CSL Report R-346, April 1967, Box 2, CBI PLATO Collection, 10. Although the authors mention "work at other laboratories" here and in other publications (such as D. L. Bitzer and H. G. Slottow, "Principles and Application of the Plasma Display Panel," *Proceedings of IEEE Symposium on Micro-Electronics and Electronic System*, Section and Professional Group on Parts, Materials and Packaging, St. Louis, MO, March 1968, Box 2, CBI PLATO Collection), they do not explicitly mention the other laboratories. In the Arora et al. paper, the authors cite only "private communication" with "W. N. Mayer" and with "J. Morrison, A. Soble, and J. Marken."

57. Arora et al., "The Plasma Display Panel," 11. Although the Arora et al. paper does not mention Mayer's laboratory affiliation, Mayer and others, as employees of Control Data Corporation, filed a patent in 1968 for a "Gaseous Display Control." See United States Patent 3,573,542.

58. Donald Bitzer and Hiram Slottow, United States Patent 3,601,532.

59. Donald Bitzer and H. G. Slottow, "The Plasma Display Panel—A Digitally Addressable Display with Inherent Memory," in *Proceedings of the Fall Joint Computer Conference*, November 1966, Box 2, CBI PLATO Collection; Arora et al., "The Plasma Display Panel"; Bitzer and Slottow, "Principles and Application of the Plasma Display Panel."

60. Bitzer and Slottow, "The Plasma Display Panel," 546.

61. Bitzer and Slottow, "Principles and Application of the Plasma Display Panel," 5.

62. Bitzer and Slottow, "The Plasma Display Panel," 545.

63. "The need for a full graphics display . . ." in Larry F. Weber, "History of the Plasma Panel Display," *IEEE Transactions on Plasma Science* 34, no. 2 (2006): 268; "to make it more comfortable . . ." in University Communications, "Alma Mater Inducts Bitzer," NC State News, March 23, 2011, https://news.ncsu.edu/2011/03/alma-mater-inducts-bitzer/.

64. Bitzer and Slottow, "Principles and Application of the Plasma Display Panel."

65. "Golden Age of Education" from Andrew R. Molnar, oral history conducted in 1991 by William Aspray, IEEE History Center, Hoboken, NJ, http://ethw.org/Oral-History:Andrew_R._Molnar; see also Andrew Molnar, oral

history interview by William Aspray, Charles Babbage Institute, retrieved from the University of Minnesota Digital Conservancy, http://hdl.handle.net/11299/107509.

66. Wayne J. Urban and Jennings L. Wagoner Jr., *American Education: A History*, 4th ed. (New York: Routledge, 2009).

67. John L. Rudolph, *Scientists in the Classroom: The Cold War Reconstruction of American Science Education* (New York: Palgrave, 2002).

68. Christopher J. Phillips, *The New Math: A Political History* (Chicago: University of Chicago Press, 2015); Jamie Cohen-Cole, *The Open Mind: Cold War Politics and the Sciences of Human Nature* (Chicago: University of Chicago Press, 2014).

69. Rudolph, *Scientists in the Classroom*, 2.

70. United States President's Science Advisory Committee, Panel on Computers in Education (John R. Pierce et al.), *Computers in Higher Education: Report of the President's Science Advisory Committee* (Washington, DC: The White House, February 1967).

71. Lyndon B. Johnson, "Special Message to the Congress: Education and Health in America," February 28, 1967, text online by Gerhard Peters and John T. Woolley, *The American Presidency Project*, http://www.presidency.ucsb.edu/ws/?pid=28668.

72. Molnar, Charles Babbage Institute interview.

73. Bitzer and Slottow, "The Plasma Display Panel," 544.

74. PLATO Progress Report for September 1966–February 1967, Box 22, CBI PLATO Collection, 127; emphasis added.

75. Ibid., 127–128.

76. Ibid.; Elisabeth R. Lyman, "A Descriptive List of PLATO Programs 1960–1970, 2nd ed. Revised," May 1970, http://eric.ed.gov/?id=ED042623.

77. Lyman, "A Descriptive List of PLATO Programs 1960–1970"; Avner and Tenczar, *The TUTOR Manual*.

78. Daniel Alpert and Donald Bitzer, "Advances in Computer-Based Education: A Progress Report on the PLATO Program," July 1969, http://eric.ed.gov/?id=ED124133, 1.

79. Lyman, "A Descriptive List of PLATO Programs 1960–1970"; Lyman, "PLATO Highlights, 3rd Rev."; Stewart A. Denenberg, "A Personal Evaluation of the PLATO System," *SIGCUE Outlook* 12, no. 2 (April 1978): 3–10.

80. Lyman, "PLATO Highlights, 3rd Rev."; Jack Stifle, "A Plasma Display Terminal," CERL Report X-15, March 1970, Box 4, CBI PLATO Collection.

81. Lyman, "PLATO Highlights, 3rd Rev."

82. Allen L. Hammond, "Computer-Assisted Instruction: Two Major Demonstrations," *Science* 176, no. 4039 (June 9, 1972): 1110–1112.

83. Lyman, "PLATO Highlights, 3rd Rev."

84. David V. Meller, "Using PLATO IV," Education Resources Information Center, October 1975, http://eric.ed.gov/?id=ED124144, 1.1.

85. Ibid.

86. Bruce Sherwood, "Status of PLATO IV," *SIGCUE Outlook* 6, no. 3 (June 1972): 3–6, doi:10.1145/965887.965888.

87. F. A. Ebeling, R. S. Goldhor, and R. L. Johnson, "A Scanned Infrared Light Beam Touch Entry System," *Proceedings of Society of Information Displays*,

1972, Box 6, CBI PLATO Collection; Donald L. Bitzer, "Computer Assisted Education," *Theory into Practice* 12, no. 3 (June 1, 1973): 173–178.

88. Sherwood, "Status of PLATO IV."

89. Meller, "Using PLATO IV."

90. Ibid., 1.1.

91. Ibid., 1.3.

92. Ibid., 1.6.

93. Ibid., 1.12.

94. Ibid., 2.1.

95. Cynthia Solomon, *Computer Environments for Children: A Reflection on Theories of Learning and Education* (Cambridge, MA: MIT Press, 1988).

96. Amy Fahey, "Educational Technology Timeline, Early 1960s and Early 2000s, CI335-CTER2 Assignment 8, PLATO Instructional Computing System," archived at perma.cc/AFN9-EJLT.

97. The most comprehensive work on Engelbart is Thierry Bardini, *Bootstrapping: Douglas Engelbart, Coevolution, and the Origins of Personal Computing* (Stanford, CA: Stanford University Press, 2000). The demonstration is viewable at https://www.youtube.com/watch?v=yJDv-zdhzMY. Many works describe it, including Howard Rheingold, *Tools for Thought: The People and Ideas behind the Next Computer Revolution* (New York: Simon & Schuster, 1985); and John Markoff, *What the Dormouse Said: How the Sixties Counterculture Shaped the Personal Computer Industry* (New York: Viking, 2005).

98. Alan Kay, "The Early History of Smalltalk," in *History of Programming Languages II*, ed. Thomas Bergin Jr. and Richard Gibson Jr. (New York: ACM, 1996), 511–579.

99. Michael A. Hiltzik, *Dealers of Lightning: Xerox PARC and the Dawn of the Computer Age* (New York: HarperBusiness, 1999); Douglas K. Smith and Robert C. Alexander, *Fumbling the Future: How Xerox Invented, and Then Ignored, the First Personal Computer* (New York: W. Morrow, 1988).

100. Steven Levy, *Insanely Great: The Life and Times of Macintosh, the Computer That Changed Everything* (New York: Penguin, 2000), "mother of all demos" on 42, PARC quote on 51.

101. Malcolm Gladwell, "Creation Myth: Xerox PARC, Apple, and the Truth about Innovation," *New Yorker*, May 16, 2011, http://www.newyorker.com/magazine/2011/05/16/creation-myth.

102. Alex Soojung-Kim Pang, *Making the Macintosh: Technology and Culture in Silicon Valley*, online project documenting the history of the Macintosh, http://web.stanford.edu/dept/SUL/library/mac/index.html; Jobs's visit to PARC addressed in "The Xerox PARC Visit," http://web.stanford.edu/dept/SUL/library/mac/parc.html, archived at perma.cc/48DZ-Z8ZE.

103. Gladwell, "Creation Myth"; emphasis in original.

7 PLATO's Republic (or, the Other ARPANET)

1. PLATO Quarterly Progress Report for September–November 1962, Box 22, Collection CBI 133: University of Illinois at Urbana-Champaign, Computer-Based Education Research Laboratory PLATO Reports, PLATO

Documents and CERL Progress Reports (hereafter CBI PLATO Collection), Charles Babbage Institute, Minneapolis, Minnesota.

2. PLATO Quarterly Progress Report for December 1962–February 1963, Box 22, CBI PLATO Collection, 27.

3. PLATO Quarterly Progress Report for December 1963–February 1964, Box 22, CBI PLATO Collection, 2.

4. R. A. Avner and Paul Tenczar, *The TUTOR Manual*, CERL Report X-4, January 1969, Box 3, CBI PLATO Collection.

5. PLATO Quarterly Progress Report for September–November 1964, Box 22, CBI PLATO Collection, 41–42.

6. PLATO Quarterly Progress Report for December 1964–February 1965, Box 22, CBI PLATO Collection.

7. For an introduction to the rights and protest movements of the long 1960s, see Terry Anderson, *The Movement and the Sixties: Protest in America from Greensboro to Wounded Knee* (New York: Oxford University Press, 1995).

8. Stuart Umpleby, "Citizen Sampling Simulations: A Method for Involving the Public in Social Planning," *Policy Sciences* 1, no. 1 (1970): 373.

9. PLATO Quarterly Progress Report for September–November 1965, Box 22, CBI PLATO Collection.

10. Charles Osgood and Stuart Umpleby, *A Computer-Based System for Exploration of Possible Futures for Mankind 2000*, Progress Report for Mankind 2000, August 1967, Box 2, CBI PLATO Collection.

11. Valarie Lamont, ed., *The Alternative Futures Project at the University of Illinois Newsletter*, no. 1 (January 1971): 1, http://www.gwu.edu/~umpleby/project_newsletters.html, archived to https://perma.cc/9NYS-PDZ9; UIUC Public Information Office, "Immediate Release Mailed 3 / 28 / 69" (on futures research), http://www.gwu.edu/~umpleby/afp/U%20of%20I%20Press%20Release%203_28_1969.pdf, archived to https://perma.cc/528Q-9FBU.

12. UIUC Public Information Office, "Immediate Release Mailed 3 / 28 / 69"; RAND Corporation, "History and Mission," http://www.rand.org/about/history.html, archived to https://perma.cc/ELW9-N9BM.

13. Umpleby, "Citizen Sampling Simulations," 361.

14. Ibid., 363.

15. Stuart Umpleby, *The Teaching Computer as a Device for Social Science Research*, Computer-based Education Research Laboratory (CERL) Report X-7, May 1969, Box 4, CBI PLATO Collection.

16. Ibid., 6.

17. Ibid.

18. Valarie Lamont, "PLATO Program on the Boneyard Creek," Institute of Communications Research, University of Illinois, Urbana, June 1970, Box 4, CBI PLATO Collection.

19. The Concerned Engineers for the Restoration of the Boneyard (CERB), "The Restoration of the Boneyard: A Paper Presented to the Staff of General Engineering 242, University of Illinois, in Partial Fulfillment of the Requirements for the Degree of Bachelor of Science in General Engineering," January 1970, Illinois Harvest Digital Collections, http://hdl.handle.net/10111/UIUCBONEYARD:6261448_1 (hereafter CERB Report).

20. Valarie Lamont, *New Directions for the Teaching Computer: Citizen Participation in Community Planning*, CERL Report X-34, Computer-Based Education Research Laboratory, University of Illinois, Urbana, Illinois, July 1972, Box 5, CBI PLATO Collection.

21. Ibid.

22. Ibid., 4–5.

23. The historian Adam Rome argues that the first Earth Day of 1970 did not mark the beginnings of local environmental activism around the United States but rather a concentrated expression of that activism on a national level. Adam Rome, *The Genius of Earth Day: How a 1970 Teach-In Unexpectedly Made the First Green Generation* (New York: Hill and Wang, 2013).

24. Lamont, *New Directions for the Teaching Computer*, 8. Here I am invoking the multiple meanings of "demonstration"—one could "demonstrate" a computing system, like PLATO, but one could also call a gathering of individuals a "demonstration." The analysis I have in mind here is Elaine Tyler May's consideration of the multiple meanings of "containment" in *Homeward Bound: American Families in the Cold War Era*, fully revised and updated 20th anniversary edition (New York: Basic Books, 2008).

25. Lamont, *New Directions for the Teaching Computer*, 13; all comments reported here as they appeared in the original report (including lack of capitalization).

26. Lamont, "PLATO Program on the Boneyard Creek," slides 97–105.

27. For two works on the role of education in the civil rights movement, see Barbara Ransby, *Ella Baker and the Black Freedom Movement: A Radical Democratic Vision* (Chapel Hill: University of North Carolina Press, 2003); Katherine Charron, *Freedom's Teacher: The Life of Septima Clark* (Chapel Hill: University of North Carolina Press, 2009).

28. Lamont, "PLATO Program on the Boneyard Creek," slide 1.

29. Lamont edited the group's first newsletter for its January 1971 publication, but she noted that the group had existed since 1967 under the direction of Professor Osgood. See Valarie Lamont, ed., *The Alternative Futures Project at the University of Illinois Newsletters* 1–3, http://www.gwu.edu/~umpleby/project_newsletters.html.

30. Stuart Umpleby, *The Delphi Exploration: A Progress Report*, August 1969, Box 4, CBI PLATO Collection; Valarie Lamont and Stuart Umpleby, *Forty Information Units with Background Paragraphs for Use in a Computer-Based Exploration of the Future*, Social Implications of Science and Technology Report F-2, Institute of Communications Research, March 1970, Box 4, CBI PLATO Collection.

31. John Daleske, who created the PLATO game *Empire*, described the "association of game authors" in "PLATO—Also an Excellent Platform to Design Games," http://www.daleske.com/plato/PLATO-games.php, archived to https://perma.cc/77ZW-HBDY.

32. *Demonstration and Evaluation of the PLATO IV Computer-Based Education System (Computer-based Education for a Volunteer Armed Service Personnel Program), First Annual Report sponsored by Advanced Research Projects Agency*, ARPA

Order No. 2245, Program Code No. P3D20, U.S. Army Contract DAHC-15-73-C-0077, [Report] for the period August 1, 1972–January 1, 1974, February 1974, Box 6, CBI PLATO Collection. For additional reporting on this ARPA–PLATO collaboration, see also *Annual Report sponsored by Advanced Research Projects Agency*, ARPA Order No. 2245, Program Code No. P3D20, U.S. Army Contract DAHC-15-73-C-0077, Effective: August 1, 1972 thru June 30, 1975. Demonstration and Evaluation of the PLATO IV Computer-Based Education System (Computer-based Education for a Volunteer Armed Service Personnel Program) for the period January 1, 1974–December 31, 1974, April 1975, Box 8, CBI PLATO Collection; *Semi-Annual Report sponsored by Advanced Research Projects Agency*, ARPA Order No. 2245, Program Code No. P3D20, U.S. Army Contract DAHC-15-73-C-0077, Effective: August 1, 1972 thru June 30, 1975. Demonstration and Evaluation of the PLATO IV Computer-Based Education System (Computer-based Education for a Volunteer Armed Service Personnel Program) for the period January 1, 1975–June 30, 1975, September 1975, Box 9, CBI PLATO Collection.

33. Although ARPA changed its name to DARPA (Defense Advanced Research Projects Agency) in 1972, the PLATO documents continue to refer to ARPA.

34. *Demonstration and Evaluation of the PLATO IV Computer-Based Education System, First Annual Report.* They were (1) the U.S. Army Ordnance Center and School in Aberdeen, Maryland (Aberdeen Proving Ground); (2) the Chanute Air Force Base Technical Training Center in Champaign County, Illinois; (3) the Sheppard Air Force Base School of Health Care Sciences in Wichita County, Texas; (4) the Navy Personnel Research Development Center in San Diego, California; (5) the Naval Training Equipment Center in Orlando, Florida; (6) the U.S. Army Signal Corps and School at Fort Monmouth, New Jersey; (7) the Air Force Human Resources Laboratory at Lowry Air Force Base in Colorado; (8) the Behavioral Technologies Laboratory at the University of Southern California in Los Angeles; (9) RAND in Santa Monica, California; (10) the Human Resources Research Organization in Alexandria, Virginia; and (11) the Institute of Mathematic Studies at Stanford University in Palo Alto, California.

35. *Demonstration and Evaluation of the PLATO IV Computer-Based Education System, First Annual Report.*

36. Computer History Museum, "The Computer History Museum, SRI International, and BBN Celebrate the 40th Anniversary of First ARPANET Transmission," October 27, 2009, http://www.computerhistory.org/press/museum-celebrates-arpanet-anniversary.html.

37. Janet Abbate, *Inventing the Internet* (Cambridge, MA: MIT Press, 1999).

38. RCA Service Company quoted in Abbate, *Inventing the Internet*, 86.

39. Abbate, *Inventing the Internet*, 106.

40. *Demonstration and Evaluation of the PLATO IV Computer-Based Education System, First Annual Report*, 7.

41. Ibid.

42. All preceding quotations in this paragraph are from ibid., 9–12.

43. Bruce Sherwood and Jack Stifle, *The PLATO IV Communications System*, CERL Report X-44, June 1975, 6, Box 8, CBI PLATO Collection.

44. Abbate, *Inventing the Internet*, chapter 3.

45. The PLATO network's strength relative to ARPANET eventually became its weakness. Initially, the PLATO network worked so fluidly because all of the terminals were essentially interchangeable, and they were all connected to one mainframe computer. The connections were fast, and the identical terminals and single mainframe facilitated technical support. Moreover, PLATO had been under development for over a decade, so there was a substantial body of institutional knowledge about the system, and about using and supporting it. In contrast, the ARPANET had been developed precisely to enable connections (and resource sharing) among different computers on different systems. Over the course of the 1970s and 1980s, as minicomputers and microcomputers (or personal computers) of various makes and models proliferated at universities and corporations, the ARPANET enabled connections among all those various computers and users. The PLATO network, meanwhile, served only PLATO users.

46. *Demonstration and Evaluation of the PLATO IV Computer-Based Education System, First Annual Report*, 12.

47. Ibid., 7.

48. Meller, "Using PLATO IV," 4.1.

49. Meller, "Using PLATO IV."

50. Lyman, "PLATO Highlights. Third Rev."

51. In accordance with current discussions around protecting the identities of individuals who may not have intended to become part of the historical record when they posted online, I have used first-name-only pseudonyms for all individuals who posted to or were mentioned in these PLATO notes files.

52. ONotes 03 from Digital Surrogates from the PLATO System Notes Files, University of Illinois Archives, available as PDF or TXT files at https://archives.library.illinois.edu/e-records/index.php?dir=University%20Archives/0713010/. Please note that all subsequent references to ONotes refer to these University of Illinois Archives Digital Surrogates and will be cited as ONotes [number].

53. LessonNotes 01 from Digital Surrogates from the PLATO System Notes Files, University of Illinois Archives, available as PDF or TXT files at https://archives.library.illinois.edu/e-records/index.php?dir=University%20Archives/0713010/. Please note that all subsequent references to LessonNotes refer to these University of Illinois Archives Digital Surrogates and will be cited as LessonNotes [number].

54. LessonNotes 01 and LessonNotes 02.

55. LessonNotes 04.

56. LessonNotes 04.

57. LessonNotes 04 and LessonNotes 05.

58. LessonNotes 05.

59. LessonNotes 05.

60. LessonNotes 08.

61. ONotes 04.

62. LessonNotes 08.
63. LessonNotes 04.
64. LessonNotes 05.
65. "Some twit" in LessonNotes 01; see similar exchanges in LessonNotes 02.
66. LessonNotes 11.
67. LessonNotes 11.
68. LessonNotes 07.
69. LessonNotes 07.
70. LessonNotes 03.
71. LessonNotes 05.
72. LessonNotes 06.
73. LessonNotes 09.
74. LessonNotes 04.
75. LessonNotes 09.
76. LessonNotes 06.
77. LessonNotes 06.
78. LessonNotes 02.
79. LessonNotes 02.
80. LessonNotes 03.
81. LessonNotes 03.
82. LessonNotes 03.
83. See, for example, LessonNotes 01 through LessonNotes 05.
84. LessonNotes 06.
85. LessonNotes 06.
86. LessonNotes 06.
87. Stewart A. Denenberg, "A Personal Evaluation of the PLATO System," *SIGCUE Outlook* 12, no. 2 (1978): 3–10.
88. LessonNotes 06.
89. LessonNotes 06.
90. LessonNotes 06.
91. LessonNotes 17.
92. All quotations in paragraph from Denenberg, "A Personal Evaluation of the PLATO System," 5.
93. See, for example, numerous comments in LessonNotes 01 through LessonNotes 05.
94. LessonNotes 06.
95. LessonNotes 06.
96. For analyses of the association of teenagers with hacking during the 1980s, see Stephanie Ricker Schulte, *Cached: Decoding the Internet in Global Popular Culture* (New York: New York University Press, 2013); and Meryl Alper, "'Can Our Kids Hack It with Computers?' Constructing Youth Hackers in Family Computing Magazines (1983–1987)," *International Journal of Communication* 8 (2014): 673–698.
97. LessonNotes 10.
98. LessonNotes 13.
99. ONotes 04.

100. Biographical details about "Roger" drawn from his biography at the Computer History Museum, archived at http://perma.cc/9QPQ-UGF5. Since I have provided pseudonyms for PLATO notes users (see note 51), I am not providing a full citation here.

101. Notes involving "Denise" in ONotes 04.

102. Notes involving "Catherine" in ONotes 04.

103. "Crank call" note and responses in ONotes 18.

104. ONotes 18.

105. Notes involving "Sharon" in LessonNotes 07.

106. Jenny Korn argues that although some women who used the PLATO system during the 1970s stated that they experienced "genderlessness" on PLATO, that "genderlessness" was an example of muting: "The patriarchal culture within male-dominated computer forums has the effect not only of preventing women from talking but also of shaping and controlling women's voices." Jenny Ungbha Korn, "'Genderless' Online Discourse in the 1970s: Muted Group Theory in Early Social Computing," in *Ada's Legacy: Cultures of Computing from the Victorian to the Digital Age*, ed. Robin Hammerman and Andrew L. Russell (New York: ACM, 2016), 213–229.

107. "Peter's" comment in LessonNotes 19.

108. ONotes 18.

109. Quotations from ONotes 18; discussion of restroom incidents in ONotes 18 and ONotes 19.

110. All quotations in this paragraph are from ONotes 19.

111. Smoking in ONotes 06.

112. For example, LessonNotes 06 and LessonNotes 10. "Peter's" comment in ONotes 04.

113. "Andrew's" comment in ONotes 04; emphasis in original.

114. "Andrew's" hours in ONotes 02.

115. "Gerald's" hours in ONotes 04.

116. ONotes 19.

117. Cookie thread in ONotes 04.

118. Streaking documented with photographs in the University of Illinois Archives, as well as in the article by Patrick Wade, "Whatever Happened to: Streaking at the University of Illinois," *The News-Gazette*, January 26, 2014, archived at https://perma.cc/75LT-WG7M.

119. Discussion of PLATO streaker in ONotes 4b and ONotes 06.

120. Bill Kirkpatrick, "'It Beats Rocks and Tear Gas': Streaking and Cultural Politics in the Post-Vietnam Era," *Journal of Popular Culture* 43, no. 5 (October 2010): 1023–1047.

121. Seasonal displays in ONotes 19 and ONotes 40; quoted material in ONotes 19.

122. Quotation and Mickey Mouse thread in ONotes 18.

123. ONotes 18.

124. ONotes 18.

125. ONotes 19.

126. ONotes 19.

127. LessonNotes 06.

128. ONotes 01.

129. ONotes 02. See also Richard Shapiro, "'World Plan' Week Seeks to Promote Meditation," *Daily Illini*, January 30, 1974, 16, 18, Illinois Digital Newspaper Collections, http://idnc.library.illinois.edu/cgi-bin/illinois?a=d&d=DIL19740130&e=--------en-20--1--txt-txIN-------, archived at https://perma.cc/CE2F-K6VT.

130. ONotes 02.

131. ONotes 03.

132. ONotes 03.

133. ONotes 03.

134. Stuart A. Umpleby, "Overstatement," *Change* 6, no. 6 (July 1, 1974): 4.

135. Denenberg, "A Personal Evaluation of the PLATO System," 8.

136. Ibid.

137. LessonNotes 09.

138. Denenberg, "A Personal Evaluation of the PLATO System," 8. For more on games, see Elisabeth R. Lyman, *PLATO Curricular Materials, Number 6* (Urbana: University of Illinois Computer-Based Education Research Laboratory, 1977), https://eric.ed.gov/?id=ED151017.

139. LessonNotes 06.

140. LessonNotes 06.

141. Denenberg, "A Personal Evaluation of the PLATO System," 6.

142. ONotes 05.

143. ONotes 05.

144. Donald Lester Bitzer, "Computer Assisted Education," *Theory into Practice* 12, no. 3 (1973): 173–178.

145. Denenberg, "A Personal Evaluation of the PLATO System." See ONotes 19 for the felicitations extended to two PLATO people on their marriage.

Epilogue

1. Donald L. Bitzer, oral history interview by Sheldon Hochheiser, February 19, 1988, Charles Babbage Institute, retrieved from the University of Minnesota Digital Conservancy, http://hdl.handle.net/11299/107121.

2. PLATO Quarterly Progress Report for June–August 1961, Box 22, Collection CBI 133: University of Illinois at Urbana-Champaign, Computer-Based Education Research Laboratory PLATO Reports, PLATO Documents and CERL Progress Reports (hereafter CBI PLATO Collection), Charles Babbage Institute, Minneapolis, Minnesota.

3. Donald Bitzer, oral history interview by Mollie Price, August 17, 1982, 19, Charles Babbage Institute.

4. Bitzer, 1988 interview.

5. Ibid.

6. Ibid.; Bitzer, 1982 interview.

7. PLATO Quarterly Progress Report for September–November 1965, Box 22, CBI PLATO Collection; Bitzer, 1982 interview; Bitzer, 1988 interview. The PLATO Quarterly Progress Report for September–November 1965 (Box 22,

CBI PLATO Collection) reported the gift of the CDC computer for arrival in 1966.

8. Donald D. Bushnell, "Computer-Based Teaching Machines," *Journal of Educational Research* 55, no. 9 (June 1, 1962): 528–531; Walter Dick, "The Development and Current Status of Computer-Based Instruction," *American Educational Research Journal* 2, no. 1 (January 1, 1965): 41–54, doi:10.2307/1162068.

9. Elisabeth Van Meer, "PLATO: From Computer-Based Education to Corporate Social Responsibility," *Iterations: An Interdisciplinary Journal of Software History*, November 5, 2003, http://www.cbi.umn.edu/iterations/vanmeer.pdf.

10. Stewart A. Denenberg, "A Personal Evaluation of the PLATO System," *SIGCUE Outlook* 12, no. 2 (1978): 8.

11. For more on why CDC employed so few Illinois courses, see Van Meer, "PLATO," especially note 87.

12. Denenberg, "A Personal Evaluation of the PLATO System," 10.

13. Ibid., 7.

14. Controversy cropped up around CDC and PLATO during the late 1970s and early 1980s. In 1980, church groups castigated CDC for its sales of PLATO to South Africa under apartheid. In 1982, the *Minneapolis Tribune* challenged the veracity of CDC's advertising claims. In fact, in her article on PLATO, Elisabeth Van Meer argues that Norris attempted to market PLATO as a solution to social problems, but that veneer of corporate social responsibility was criticized by activists, investors, and journalists through controversies about truth in advertising and supporting South African apartheid. Van Meer, "PLATO."

15. *People's Computer Company* 1, no. 1 (October 1972), 1; all ellipses in original.

16. Circulation number from Fred Turner, *From Counterculture to Cyberculture: Stewart Brand, the Whole Earth Network, and the Rise of Digital Utopianism* (Chicago: University of Chicago Press, 2006), 113.

17. Steven Levy, *Hackers: Heroes of the Computer Revolution* (Sebastopol, CA: O'Reilly Media, 2010), 196–199. Levy attributes Moore's cofounding (with Gordon French) of the Homebrew Club in no small part to his frustration with Albrecht and the *People's Computer Company*. According to Levy, Albrecht was on a "planner" mission to spread computing far and wide, while Moore exhibited a "hacker" fascination with hardware. Yet Levy later notes (222) that others also dismissed the Homebrew crew as "chip-monks, people obsessed with chips."

18. *People's Computer Company* 3, no. 3 (January 1975).

19. *People's Computer Company* 3, no. 4 (March 1975), 6–7; capitalization of TINY BASIC in original.

20. Ibid.

21. John Markoff, *What the Dormouse Said: How the Sixties Counterculture Shaped the Personal Computer Industry* (New York: Viking, 2005), 264–265. Note the play on byte, overbyte, and overbite, leading to "orthodontia" in the title.

22. Jim Warren quoted in Markoff, *What the Dormouse Said*, 265.

23. Warren quoted in Levy, *Hackers*, 235.

24. Bill Gates, oral history interview by David Allison, National Museum of American History (Smithsonian Institution), http://americanhistory.si.edu/comphist/gates.htm#tc3.

25. Ibid.

26. Ibid.

27. The letter is now available on Wikimedia Commons via the DigiBarn Computer Museum, which notes that it was published not only in the *Homebrew Computer Club Newsletter* but also in *Computer Notes*, *People's Computer Company*, and *Radio-Electronics*. "File: Bill Gates Letter to Hobbyists.jpg," https://commons.wikimedia.org/wiki/File:Bill_Gates_Letter_to_Hobbyists.jpg.

28. Jim Warren, Editor of *Dr. Dobb's Journal*, April 10, 1976, http://www.vintagecomputer.net/pcc/billgatesopenletter.pdf.

29. Details on Apple from Levy, *Hackers*, 256–259.

30. From an Apple ad quoted in Levy, *Hackers*, 258.

31. Nelson quoted in Levy, *Hackers*, 274.

32. *Dataline* 6, no. 2 (November–December 1978), Minnesota Historical Society.

33. *Dataline* 5, no. 1 (September–October 1977).

34. *Dataline* 5, no. 3 (January–February 1978).

35. Brumbaugh quoted in David H. Ahl, "Interview with Ken Brumbaugh [Director of MECC Instructional Services]," *Creative Computing*, March 1981, 116.

36. Steve Jobs, oral history interview by Daniel Morrow, National Museum of American History (Smithsonian Institution), http://americanhistory.si.edu/comphist/sj1.html#import.

37. Dale Eugene LaFrenz, oral history interview by Jude E. O'Neill, April 13, 1995, Charles Babbage Institute, retrieved from the University of Minnesota Digital Conservancy, http://hdl.handle.net/11299/107423.

38. By 1981, MECC had obtained over three thousand Apple IIs for its schools, ranging from kindergartens to colleges. Scott Mace, "Minnesota's MECC Educates Next Generation of Computer Users," *InfoWorld*, December 7, 1981.

39. LaFrenz, interview.

40. Don G. Rawitsch, "Implanting the Computer in the Classroom: Minnesota's Successful Statewide Program," *Phi Delta Kappan* 62, no. 6 (1981): 453–454.

41. LaFrenz, interview.

42. Thomas J. Misa, *Digital State: The Story of Minnesota's Computing Industry* (Minneapolis: University of Minnesota Press, 2013), chapter 7.

43. Ken Brumbaugh, "Microcomputers vs. Timesharing," *Creative Computing*, March 1981, 132.

44. My interpretation of this shift from computing citizens to the "CONSUMER MARKET!"—blasted by Ted Nelson to audiences at the first Computer Faire (recounted in Levy, *Hackers*, 274)—is informed by Liz Cohen's influential argument in *A Consumers' Republic: The Politics of Mass Consumption in Postwar America* (New York: Vintage Books, 2003) that the post–World War II

promise of prosperity, symbolized by the acquisition of consumer goods and the attendant emphasis on mass consumption, transformed American citizenship.

45. Kevin Driscoll addresses the history of BBSs in "Hobbyist Internetworking and the Popular Internet Imaginary: Forgotten Histories of Networked Personal Computing, 1978–1998" (PhD diss., University of Southern California, 2014), which he is revising for a book tentatively titled *The Modem World: A Prehistory of Social Media.*

46. Virginia Heffernan, *Magic and Loss: The Internet as Art* (New York: Simon & Schuster, 2016).

47. Ibid., 18.

48. Ibid., 21.

49. I thank Matthew Kirschenbaum for calling my attention to this concept in his excellent book *Track Changes: A Literary History of Word Processing* (Cambridge, MA: Harvard University Press, 2016).

Bibliography

Archives and Manuscript Collections

ANDOVER, MASSACHUSETTS

Phillips Academy
- John M. Kemper Records
- *The Phillipian* [student newspaper] Collection
- Richard Pieters Information File
- Theodore R. Sizer Records

BROOKLYN, NEW YORK

Bern Dibner Library, New York University Tandon School of Engineering
- *Polytechnic Reporter* Collection

HANOVER, NEW HAMPSHIRE

Dartmouth College Rauner Special Collections Library
- Dartmouth College History Collection
- Information File for the Computation Center
- Information File for Time-Sharing
- Oral history interview with John G. Kemeny, 1984
- Oral history interview with Thomas E. Kurtz, 2002
- Papers of John G. Kemeny
- Papers of Thomas E. Kurtz
- Papers of Sidney Marshall
- Records of Dartmouth College Computing Services
- Stephen Garland Papers

MINNEAPOLIS, MINNESOTA

Charles Babbage Institute, University of Minnesota
- Control Data Corporation Records of Marketing, Sales, and Public Relations
- Control Data Corporation Records of Newspaper and Magazine Articles
- Control Data Corporation Records on Computer-Based Education

EDUCOM Survey Records
Oral history interview with Donald Lester Bitzer, 1982
Oral history interview with Donald Lester Bitzer, 1988
Oral history interview with Fernando Corbató
Oral history interview with Robert M. Fano
Oral history interview with Dale Eugene LaFrenz
Oral history interview with Andrew Molnar
Total Information for Educational Systems (TIES) Records
University of Illinois at Urbana-Champaign, Computer-Based Education Research Laboratory PLATO Reports, PLATO Documents and CERL Progress Reports
University of Minnesota Archives
Information File for Computer Center
Information File for University High School
Marvin L. Stein Papers
Office of the President Records
Office of the Vice President for Academic Administration Records
University High School Records
Walker Art Center Archives
New Learning Spaces and Places Exhibition Records

MOUNTAIN VIEW, CALIFORNIA

Computer History Museum
Still Images Collection

PALO ALTO, CALIFORNIA

Stanford University Special Collections and University Archives
Liza Loop Papers
People's Computer Company Digital Collection

SAINT PAUL, MINNESOTA

Minnesota Historical Society
Microfilm Collection of *Dataline*
Microfilm Collection of *Systems Update*
Minnesota Educational Computing Consortium (MECC) Records

URBANA, ILLINOIS

University of Illinois at Urbana-Champaign
H. Gene Slottow Papers
Photographic Subject File
PLATO System Notes Files, 1972–1976 (Digital Surrogates)
Public Information Subject File

WASHINGTON, DC

The National Museum of American History
Oral history interview with Bill Gates
Oral history interview with Steve Jobs

WINDSOR, CONNECTICUT

The Loomis Chaffee School Archives
Loomis Bulletin Collection

Interviews and Inquiries Conducted by the Author

Robert (Bob) Albrecht, electronic mail, January 2014
Medford Cashion III, electronic mail, December 2013
Alexander Conn, electronic mail, December 2013
Jere R. Daniell, Hanover, New Hampshire, October 2011
Greg Dobbs, electronic mail, September 2013
Linda Borry Hausmann, Minneapolis, Minnesota, May 2013
Anthony Knapp, electronic mail, July 2013
Dale LaFrenz, Minneapolis, Minnesota, May 2013
James Lawrie, electronic mail, August 2013
David Magill, electronic mail, January 2014
Francis Marzoni, electronic mail, December 2013
David Ziegler, electronic mail, December 2013

Documents from the Education Resources Information Center (ERIC)

Alpert, Daniel, and Donald Bitzer. *Advances in Computer-Based Education: A Progress Report on the PLATO Program*. July 1969. http://eric.ed.gov/?id=ED124133.

Braun, Ludwig, and James Friedland. "Huntington II Simulation Program—BUFLO. Student Workbook, Teacher's Guide, and Resource Handbook." March 1974. http://eric.ed.gov/?id=ED179411.

Haugo, John E. "Minnesota Educational Computing Consortium." April 1973. http://eric.ed.gov/?id=ED087434.

Kemeny, John G., and Thomas E. Kurtz. *The Dartmouth Time-Sharing Computing System. Final Report*. June 1967. http://eric.ed.gov/?id=ED024602.

Kurtz, Thomas E. *Demonstration and Experimentation in Computer Training and Use in Secondary Schools, Activities and Accomplishments of the First Year*. October 1968. http://eric.ed.gov/?id=ED027225.

Lyman, Elisabeth R. "A Descriptive List of PLATO Programs 1960–1970, 2nd ed. Revised." May 1970. http://eric.ed.gov/?id=ED042623.

———. "PLATO Highlights. Third Revision." December 1975. http://eric.ed.gov/?id=ED124143.

Meller, David V. "Using PLATO IV." October 1975. http://eric.ed.gov/?id=ED124144.

Molnar, Andrew R. "The Computer and the Fourth Revolution." Paper presented at the Association for Educational Data Systems Annual Convention, New Orleans, LA, April 16–19, 1973. http://files.eric.ed.gov/fulltext/ED087417.pdf.

Sherwood, Bruce Arne, and Jack Stifle. "The PLATO IV Communications System." June 1975. http://eric.ed.gov/?id=ED124148.

Sydow, James Armin. “Computers in Physical Education.” November 1974. http://eric.ed.gov/?id=ED096972.
———. “Secondary Athletics and the Computer.” January 1974. http://eric.ed.gov/?id=ED096971.
Visich, Marian, and Ludwig Braun. “The Use of Computer Simulations in High School Curricula.” January 1974. http://eric.ed.gov/?id=ED089740.

Nonarchival Sources

Abbate, Janet. “Government, Business, and the Making of the Internet.” *Business History Review* 75, no. 1 (2001): 147–176.
———. *Inventing the Internet.* Cambridge, MA: MIT Press, 1999.
———. *Recoding Gender: Women's Changing Participation in Computing.* Cambridge, MA: MIT Press, 2012.
Abramson, Norman. “Digital Broadcasting in Hawaii—The ALOHA System.” *EDUCOM: Bulletin of the Interuniversity Communications Council* 9, no. 1 (Spring 1974): 9–13.
Ahl, David. “Interview with Ken Brumbaugh [Director of MECC Instructional Services].” *Creative Computing*, March 1981, 116.
Akera, Atsushi. *Calculating a Natural World: Scientists, Engineers, and Computers during the Rise of U.S. Cold War Research.* Cambridge, MA: MIT Press, 2007.
———. “Voluntarism and the Fruits of Collaboration: The IBM User Group, Share.” *Technology and Culture* 42, no. 4 (October 1, 2001): 710–736.
Alberts, Gerard, and Ruth Oldenziel, eds. *Hacking Europe: From Computer Culture to Demoscenes.* London: Springer-Verlag, 2014.
Albrecht, Bob. *My Computer Likes Me When I Speak in BASIC.* Menlo Park, CA: Dymax, 1972.
Allen, Danielle, and Jennifer S. Light, eds. *From Voice to Influence: Understanding Citizenship in a Digital Age.* Chicago: University of Chicago Press, 2015.
Alper, Meryl. “‘Can Our Kids Hack It with Computers?’ Constructing Youth Hackers in Family Computing Magazines (1983–1987).” *International Journal of Communication* 8 (2014): 673–698.
Alpert, Daniel, and Donald Bitzer. “Advances in Computer-Based Education.” *Science* 167, no. 3925 (1970): 1582–1590.
Anderson, Bethany. “The Birth of the Computer Age at Illinois.” University of Illinois Archives (2013). http://archives.library.illinois.edu/blog/birth-of-the-computer-age. Archived at perma.cc/RJ57–9NV8.
Anderson, Terry H. *The Movement and the Sixties: Protest in America from Greensboro to Wounded Knee.* New York: Oxford University Press, 1995.
Aspray, William. *Computing before Computers.* Ames: Iowa State University Press, 1990.
———. *John von Neumann and the Origins of Modern Computing.* Cambridge, MA: MIT Press, 1991.
Aspray, William, and Paul E. Ceruzzi, eds. *The Internet and American Business.* Cambridge, MA: MIT Press, 2008.
Aspray, William, and Jeffrey Yost. “New Voices, New Topics.” *IEEE Annals of the History of Computing* 33, no. 2 (2011): 4–8.

Bailey, Beth. *Sex in the Heartland.* Cambridge, MA: Harvard University Press, 2002.

Baran, Paul. "The Future Computer Utility." In *The Computer Impact*, edited by Irene Taviss, 81–92. Englewood Cliffs, NJ: Prentice-Hall, 1970.

Bardini, Thierry. *Bootstrapping: Douglas Engelbart, Coevolution, and the Origins of Personal Computing.* Stanford, CA: Stanford University Press, 2000.

Bitzer, Donald Lester. "Computer Assisted Education." *Theory into Practice* 12, no. 3 (1973): 173–178.

———. "Signal Amplitude Limiting and Phase Quantization in Antenna Systems." PhD diss., University of Illinois at Urbana-Champaign, 1960.

Bitzer, Donald Lester, Peter Braunfeld, and Wayne Lichtenberger. "PLATO: An Automatic Teaching Device." *IRE Transactions on Education* E-4 (December 1961): 157–161.

Bitzer, Donald Lester, Elisabeth R. Lyman, and John A. Easley. "The Uses of PLATO: A Computer Controlled Teaching System." In *Proceedings of the 1966 Clinic on Library Applications of Data Processing*, edited by Herbert Goldhor, 34–46. Urbana, IL: Graduate School of Library Science, 1966.

Bitzer, Maryann D., and Donald L. Bitzer. "Teaching Nursing by Computer: An Evaluative Study." *Computers in Biology and Medicine* 3, no. 3 (1973): 187–204.

Boenig-Liptsin, Margarita. "Making Citizens of the Information Age: A Comparative Study of the First Computer Literacy Programs for Children in the United States, France, and the Soviet Union, 1970–1990." PhD diss., Harvard University, 2015.

Bolter, Jay David, and Richard Grusin. *Remediation: Understanding New Media.* Cambridge, MA: MIT Press, 1998.

Bourin, Marc, and Alex Conn. "Aspects de l'enseignement de l'informatique aux États-Unis." *Arts et Manufactures* 190 (October 1968): 43–46.

boyd, danah. *It's Complicated: The Social Lives of Networked Teens.* New Haven, CT: Yale University Press, 2015.

Brand, Stewart. "SPACEWAR: Fanatic Life and Symbolic Death among the Computer Bums." *Rolling Stone*, December 7, 1972.

Braun, Ludwig. "Acceptance of Computers in Education." *SIGCUE Outlook* 5, no. 4 (August 1971): 171–174.

———. "BASIC Is Alive and—Well?" *SIGCUE Outlook* 5, no. 5 (October 1971): 217–220.

Braunfeld, P. G., and L. D. Fosdick. "The Use of an Automatic Computer System in Teaching." *IRE Transactions on Education* 5, no. 3 (September 1962): 156–167.

Broadhead, Nancy. "Consulting Is Not a Matter of Facts." In *SIGUCCS '81: Proceedings of the 9th Annual Conference on User Services.* New York: ACM, 1981.

———. "The Unique Advantages of the Macintosh." In *SIGUCCS '85: Proceedings of the 13th Annual Conference on User Services: Pulling It All Together.* New York: ACM, 1985.

Brumbaugh, Ken. "Microcomputers vs. Timesharing." *Creative Computing* (March 1981): 132.

Bushnell, Donald D. "Computer-Based Teaching Machines." *Journal of Educational Research* 55, no. 9 (June 1, 1962): 528–531.

Campbell-Kelly, Martin. *From Airline Reservations to Sonic the Hedgehog: A History of the Software Industry.* Cambridge, MA: MIT Press, 2003.

Campbell-Kelly, Martin, William Aspray, Nathan Ensmenger, and Jeffrey R. Yost. *Computer: A History of the Information Machine.* 3rd ed. Boulder, CO: Westview Press, 2014.

Campbell-Kelly, Martin, and Daniel D. Garcia-Swartz. "Economic Perspectives on the History of the Computer Time-Sharing Industry, 1965–1985." *IEEE Annals of the History of Computing* 30, no. 1 (January 2008): 16–36.

———. *From Mainframes to Smartphones: A History of the International Computer Industry.* Cambridge, MA: Harvard University Press, 2015.

———. "The History of the Internet: The Missing Narratives." *SSRN eLibrary*, December 2, 2005. http://papers.ssrn.com/sol3/papers.cfm?abstract_id=867087.

Canaday, Margot. *The Straight State: Sexuality and Citizenship in Twentieth-Century America.* Princeton, NJ: Princeton University Press, 2009.

Cassell, Justine, and Henry Jenkins, eds. *From Barbie® to Mortal Kombat: Gender and Computer Games.* Cambridge, MA: MIT Press, 2000.

Ceruzzi, Paul E. "From Scientific Instrument to Everyday Appliance: The Emergence of Personal Computers, 1970–77." *History and Technology* 13, no. 1 (1996): 1–31.

———. *A History of Modern Computing.* 2nd ed. Cambridge, MA: MIT Press, 2003.

———. *Internet Alley: High Technology in Tysons Corner, 1945–2005.* Cambridge, MA: MIT Press, 2008.

———. *Reckoners: The Prehistory of the Digital Computer, from Relays to the Stored Program Concept, 1935–1945.* Westport, CT: Greenwood Press, 1983.

Charron, Katherine. *Freedom's Teacher: The Life of Septima Clark.* Chapel Hill: University of North Carolina Press, 2009.

Cohen, Lizabeth. *A Consumer's Republic: The Politics of Mass Consumption in Postwar America.* New York: Vintage Books, 2003.

Cohen-Cole, Jamie. *The Open Mind: Cold War Politics and the Sciences of Human Nature.* Chicago: University of Chicago Press, 2014.

The Concerned Engineers for the Restoration of the Boneyard (CERB). "The Restoration of the Boneyard: A Paper Presented to the Staff of General Engineering 242, University of Illinois, in Partial Fulfillment of the Requirements for the Degree of Bachelor of Science in General Engineering." January 1970. Illinois Harvest Digital Collections. http://hdl.handle.net/10111/UIUCBONEYARD:6261448_1.

Coordinated Science Laboratory. *Progress Report for 1959–1960.* Illinois Digital Environment for Access to Learning and Scholarship. http://hdl.handle.net/2142/28161.

Corbató, Fernando J., Marjorie Merwin-Daggett, and Robert C. Daley. "An Experimental Time-Sharing System." In *Proceedings of the May 1–3, 1962, Spring Joint Computer Conference, AIEE-IRE '62 (Spring)*, 335–344. New York: ACM, 1962. doi:10.1145/1460833.1460871.

Corbató, Fernando J., Jerome H. Saltzer, and Chris T. Clingen. "Multics: The First Seven Years." In *Proceedings of the May 16–18, 1972, Spring Joint Computer Conference*, 571–583. New York: ACM, 1972.

Cornew, Ronald W., and Philip M. Morse. "Distributive Computer Networking: Making It Work on a Regional Basis." *Science* 189, no. 4202 (1975): 523–531.

Cortada, James W. *Before the Computer: IBM, NCR, Burroughs, and Remington Rand and the Industry They Created, 1865–1956*. Princeton, NJ: Princeton University Press, 1993.

Cowan, Ruth Schwartz. *More Work for Mother: The Ironies of Household Technology from the Open Hearth to the Microwave*. New York: Basic Books, 1983.

Danver, Jean. "Publication of Computer Curriculum Materials at Hewlett Packard Company." In *Computing and the Decision Makers: Where Does Computing Fit in Institutional Priorities?* Proceedings of the EDUCOM Spring Conference, April 17–18, 1974, 228–232. Princeton, NJ: EDUCOM, 1974.

Danver, Jean, and John M. Nevison. "Secondary School Use of the Time-Shared Computer at Dartmouth College." In *Proceedings of the May 14–16, 1969 Spring Joint Computer Conference*, 681–689, AFIPS. Boston: ACM, 1969.

Denenberg, Stewart A. "A Personal Evaluation of the PLATO System." *SIGCUE Outlook* 12, no. 2 (1978): 3–10.

Dick, Walter. "The Development and Current Status of Computer-Based Instruction." *American Educational Research Journal* 2, no. 1 (January 1, 1965): 41–54.

Driscoll, Kevin. "Hobbyist Inter-networking and the Popular Internet Imaginary: Forgotten Histories of Networked Personal Computing, 1978–1998." PhD diss., University of Southern California, 2014.

Eames, Charles, and Ray Eames. *A Computer Perspective*. Edited by Glen Fleck. Cambridge, MA: Harvard University Press, 1973.

Educational Research and Development Council of the Twin Cities Metropolitan Area. *Prospectus TIES: Total Information for Educational Systems*. Minneapolis, MN: Educational Research and Development Council, 1967.

Edwards, Paul. *The Closed World: Computers and the Politics of Discourse in Cold War America*. Cambridge, MA: MIT Press, 1996.

Ensmenger, Nathan. "'Beards, Sandals, and Other Signs of Rugged Individualism': Masculine Culture within the Computing Professions." *Osiris* 30 (2015): 38–65.

———. *The Computer Boys Take Over: Computers, Programmers, and the Politics of Technical Expertise*. Cambridge, MA: MIT Press, 2010.

———. "Power to the People: Toward a Social History of Computing." *IEEE Annals of the History of Computing* 26, no. 1 (2004): 96, 94–95.

Fahey, Amy. "Educational Technology Timeline, Early 1960s and Early 2000s, CI335-CTER2 Assignment 8, PLATO Instructional Computing System." Archived at perma.cc/AFN9-EJLT.

Fano, Robert M. "The MAC System: The Computer Utility Approach." *IEEE Spectrum* 2, no. 1 (1965): 56–64.

Fano, Robert M., and Fernando J. Corbató. "Time-Sharing on Computers." *Scientific American* 215, no. 3 (September 1966): 129–140.

Freeman, Jo, and Victoria Johnson, eds. *Waves of Protest: Social Movements since the Sixties.* Lanham, MD: Rowman & Littlefield, 1999.

Garvey, Anna. "The Oregon Trail Generation: Life before and after Mainstream Tech." *Social Media Week*, April 21, 2015. https://socialmediaweek.org/blog/2015/04/oregon-trail-generation.

Gazzard, Alison. *Now the Chips Are Down: The BBC Micro.* Cambridge, MA: MIT Press, 2016.

Gladwell, Malcolm. "Creation Myth: Xerox PARC, Apple, and the Truth about Innovation." *New Yorker*, May 16, 2011. http://www.newyorker.com/magazine/2011/05/16/creation-myth.

Goldin, Claudia, Lawrence F. Katz, and Ilyana Kuziemko. "The Homecoming of American College Women: The Reversal of the Gender Gap in College." *Journal of Economic Perspectives* 20 (2006): 133–156.

Goldstein, Gordon D., Laura A. Repass, and Barbara J. Walker. *Digital Computer Newsletter* 20, no. 1. Arlington, VA: Office of Naval Research, January 1968. http://www.dtic.mil/docs/citations/AD0694655.

Goldstein, Gordon D., and Margo A. Sass. *Digital Computer Newsletter* 17, no. 2. Arlington, VA: Office of Naval Research, April 1965. http://www.dtic.mil/docs/citations/AD0694644.

Goodwin, Jeff, and James M. Jasper, eds. *The Social Movements Reader: Cases and Concepts.* Malden, MA: Blackwell Publishing, 2003.

Greenberger, Martin, ed. *Computers and the World of the Future.* Cambridge, MA: MIT Press, 1962.

———, ed. *Computers, Communications, and the Public Interest.* Baltimore: Johns Hopkins Press, 1971.

———. "The Computers of Tomorrow." *Atlantic Monthly*, May 1964. https://www.theatlantic.com/past/docs/unbound/flashbks/computer/greenbf.htm.

Greenberger, Martin, Julius Aronofsky, James L. McKenney, and William F. Massy. "Computer and Information Networks." *Science* 182, no. 4107 (October 5, 1973): 29–35.

———, eds. *Networks for Research and Education: Sharing Computing and Information Resources Nationwide.* Cambridge, MA: MIT Press, 1974.

Hammond, Allen L. "Computer-Assisted Instruction: Two Major Demonstrations." *Science* 176, no. 4039 (1972): 1110–1112.

Harkness, Deborah E. *The Jewel House: Elizabethan London and the Scientific Revolution.* New Haven, CT: Yale University Press, 2007.

Hauben, Michael, and Ronda Hauben. *Netizens: On the History and Impact of Usenet and the Internet.* Los Alamitos, CA: IEEE Computer Society Press, 1997.

Heffernan, Virginia. *Magic and Loss: The Internet as Art.* New York: Simon & Schuster, 2016.

Hicks, Marie. *Programmed Inequality: How Britain Discarded Women Technologists and Lost Its Edge in Computing.* Cambridge, MA: MIT Press, 2017.

Hiltzik, Michael A. *Dealers of Lightning: Xerox PARC and the Dawn of the Computer Age.* New York: HarperBusiness, 1999.

Hodge, Carle, ed. "EDUCOM Means Communication." *EDUCOM: Bulletin of the Interuniversity Communications Council* 1, no. 1 (January 1966): inside cover.

———, ed. "Networks for Instant Information." *EDUCOM: Bulletin of the Interuniversity Communications Council* 1, no. 7 (October 1966): 3–8.

Holznagel, Donald, Dale E. LaFrenz, and Michael Skow. "Historical Development of Minnesota's Instructional Computing Network." In *Proceedings of the 1975 Annual Conference*, 79–80. New York: ACM, 1975. doi:10.1145/800181.810282.

Horowitz, Roger, and Arwen Mohun, eds. *His and Hers: Gender, Consumption, and Technology.* Charlottesville: University Press of Virginia, 1998.

Hughes, Thomas Parke. *Networks of Power: Electrification in Western Society, 1880–1930.* Baltimore: Johns Hopkins University Press, 1993.

Hunter, Beverly. *Learning Alternatives in U.S. Education: Where Student and Computer Meet.* Englewood Cliffs, NJ: Educational Technology Publications, 1975.

Irwin, Alan. "Citizen Science and Scientific Citizenship: Same Words, Different Meanings?" In *Science Communication Today—2015*, edited by B. Schiele, J. L. Marec, and P. Baranger, 29–38. Nancy, France: Presses Universitaires de Nancy, 2015.

———. "Constructing the Scientific Citizen: Science and Democracy in the Biosciences." *Public Understanding of Science* 10, no. 1 (2001): 1–18.

Isaacson, Walter. *The Innovators: How a Group of Hackers, Geniuses, and Geeks Created the Digital Revolution.* New York: Simon & Schuster, 2015.

Jancer, Matt. "How You Wound Up Playing *The Oregon Trail* in Computer Class." *Smithsonian Magazine Online*, July 22, 2016. http://www.smithsonianmag.com/innovation/how-you-wound-playing-em-oregon-trailem-computer-class-180959851.

Jenkins, Evan. "The Potential of PLATO." *Change* 8, no. 2 (1976): 6–9.

Jepsen, Thomas C. *My Sisters Telegraphic: Women in the Telegraph Office, 1846–1950.* Athens: Ohio University Press, 2000.

Johnson, David. *The Lavender Scare: The Cold War Persecution of Gays and Lesbians in the Federal Government.* Chicago: University of Chicago Press, 2004.

Johnson, Lyndon B. "Special Message to the Congress: Education and Health in America," February 28, 1967. Text online by Gerhard Peters and John T. Woolley. *The American Presidency Project.* http://www.presidency.ucsb.edu/ws/?pid=28668.

Johnstone, Bob. *Never Mind the Laptops: Kids, Computers, and the Transformation of Learning.* New York: iUniverse, 2003.

Kafai, Yasmin B., Carrie Heeter, Jill Denner, and Jennifer Y. Sun, eds. *Beyond Barbie® and Mortal Kombat: New Perspectives on Gender and Gaming.* Cambridge, MA: MIT Press, 2008.

Katz, Barry M. *Make It New: The History of Silicon Valley Design.* Cambridge, MA: MIT Press, 2015.

Kay, Alan. "The Early History of Smalltalk." In *History of Programming Languages II*, edited by Thomas Bergin Jr. and Richard Gibson Jr., 511–579. New York: ACM, 1996.

Kemeny, John G. *Man and the Computer.* New York: Scribner, 1972.

———. "The Question of Networks: What Kind and Why?" *EDUCOM: Bulletin of the Interuniversity Communications Council* 8, no. 2 (Summer 1973): 18–21.

———. *Random Essays on Mathematics, Education and Computers.* Englewood Cliffs, NJ: Prentice-Hall, 1964.

Kemeny, John G., and Thomas E. Kurtz. *Back to BASIC: The History, Corruption, and Future of the Language.* Reading, MA: Addison-Wesley, 1985.

———. *BASIC Programming.* New York: John Wiley & Sons, 1967.

———. "Dartmouth Time-Sharing." *Science* 162, no. 3850 (October 11, 1968): 223–228.

Kessler, H. Eugene. "Education, Computing and Networking: A Review and Prospect." *EDUCOM: Bulletin of the Interuniversity Communications Council* 9, no. 1 (Spring 1974): 2–8.

Kirkpatrick, Bill. "'It Beats Rocks and Tear Gas': Streaking and Cultural Politics in the Post-Vietnam Era." *Journal of Popular Culture* 43, no. 5 (October 2010): 1023–1047.

Kirschenbaum, Matthew. *Track Changes: A Literary History of Word Processing.* Cambridge, MA: Harvard University Press, 2016.

Kleeman, Richard P. "University High Is 'Lab' for Curriculums That Are 'Far Out.'" *Minneapolis Tribune*, February 27, 1966.

Kopf, John. "TYMNET as a Multiplexed Packet Network." In *Proceedings of the June 13–16, 1977 National Computer Conference*, 609–613. New York: ACM, 1977.

Korn, Jenny Ungbha. "'Genderless' Online Discourse in the 1970s: Muted Group Theory in Early Social Computing." In *Ada's Legacy: Cultures of Computing from the Victorian to the Digital Age*, edited by Robin Hammerman and Andrew L. Russell, 213–229. New York: ACM, 2016.

Kurtz, Thomas E. "BASIC Session, Chairman: Thomas Cheatham." In *History of Programming Languages*, edited by Richard L. Wexelblat, 515–549. New York: Academic Press, 1981.

LaFrenz, Dale Eugene. "The Documentation and Formative Evaluation of a Statewide Instructional Computing Network in the State of Minnesota." PhD diss., University of Minnesota, 1978.

Lamont, Valarie, ed. *The Alternative Futures Project at the University of Illinois*, no. 1 (January 1971). http://www.gwu.edu/~umpleby/project_newsletters.html. Archived at https://perma.cc/9NYS-PDZ9.

Lécuyer, Christophe. *Making Silicon Valley: Innovation and the Growth of High Tech, 1930–1970.* Cambridge, MA: MIT Press, 2006.

Lee, J. A. N. "The Rise and Fall of the General Electric Corporation Computer Department." *IEEE Annals of the History of Computing* 17, no. 4 (Winter 1995): 24–45.

Lee, J. A. N., Robert Mario Fano, Allan L. Scherr, Fernando J. Corbató, and Victor A. Vyssotsky. "Project MAC (Time-Sharing Computing Project)." *IEEE Annals of the History of Computing* 14, no. 2 (1992): 9–13.

Leslie, Stuart W. *The Cold War and American Science: The Military-Industrial-Academic Complex at MIT and Stanford.* New York: Columbia University Press, 1993.

Levy, Steven. *Hackers: Heroes of the Computer Revolution.* Sebastopol, CA: O'Reilly Media, 2010.

———. *Insanely Great: The Life and Times of Macintosh, the Computer That Changed Everything.* New York: Penguin, 2000.

Light, Jennifer. "When Computers Were Women." *Technology and Culture* 40, no. 3 (1999): 455–483.

Loory, Stuart H. "Computer Hookup to Home Foreseen." *New York Times,* December 5, 1966.

———. "Dartmouth Expands Computer Facility." *New York Times,* December 4, 1966.

Luerhmann, Arthur W., and John M. Nevison. "Computer Use under a Free Access Policy." *Science* 184, no. 4140 (1974): 957–961.

Lussenhop, Jessica. "Oregon Trail: How Three Minnesotans Forged Its Path." *City Pages,* January 19, 2011. http://www.citypages.com/news/oregon-trail-how-three-minnesotans-forged-its-path-6745749/.

Mace, Scott. "Minnesota's MECC Educates Next Generation of Computer Users." *InfoWorld,* December 7, 1981.

Mahoney, Michael Sean. "Finding a History for Software Engineering." *IEEE Annals of the History of Computing* 26, no. 1 (2004): 8–19.

———. *Histories of Computing.* Edited by Thomas Haigh. Cambridge, MA: Harvard University Press, 2011.

———. "The Histories of Computing(s)." *Interdisciplinary Science Reviews* 30 (2005): 119–135.

Mailland, Julien, and Kevin Driscoll. *Minitel: Welcome to the Internet.* Cambridge, MA: MIT Press, 2017.

Markoff, John. *What the Dormouse Said: How the Sixties Counterculture Shaped the Personal Computer Industry.* New York: Viking, 2005.

May, Elaine Tyler. *Homeward Bound: American Families in the Cold War Era.* Fully revised and updated 20th anniversary edition. New York: Basic Books, 2008.

McCarthy, John. "Information." *Scientific American* 215, no. 3 (September 1966): 64–73.

———. "Time-Sharing Computer Systems." In *Computers and the World of the Future,* edited by Martin Greenberger, 221–236. Cambridge, MA: MIT Press, 1962.

McDonald, Christopher. "Building the Information Society: A History of Computing as a Mass Medium." PhD diss., Princeton University, 2011.

Meacham, Scott, and Joseph Mehling. *Dartmouth College: An Architectural Tour.* New York: Princeton Architectural Press, 2008.

Meadows, Donella H., Dennis L. Meadows, Jørgen Randers, and William W. Behrens. *The Limits to Growth: A Report for the Club of Rome's Project on the Predicament of Mankind.* New York: Universe Books, 1972.

Meraji, Shereen. "Fifty Years Later, 'A Better Chance' Trains Young Scholars." National Public Radio (NPR), June 9, 2013. http://www.npr.org/sections/codeswitch/2013/06/09/184798293/fifty-years-later-a-better-chance-trains-young-scholars.

Meyerowitz, Joanne. "Beyond the Feminine Mystique: A Reassessment of Postwar Mass Culture, 1946–1958." *Journal of American History* 79, no. 4 (1993): 1455–1482.

———, ed. *Not June Cleaver: Women and Gender in Postwar America, 1945–1960.* Philadelphia: Temple University Press, 1994.

Mindell, David A. *Between Human and Machine: Feedback, Control, and Computing before Cybernetics.* Baltimore: Johns Hopkins University Press, 2002.

———. *Digital Apollo: Human and Machine in Spaceflight.* Cambridge, MA: MIT Press, 2008.

Misa, Thomas J. *Digital State: The Story of Minnesota's Computing Industry.* Minneapolis: University of Minnesota Press, 2013.

———, ed. *Gender Codes: Why Women Are Leaving Computing.* Hoboken, NJ: Wiley-IEEE Computer Society, 2010.

Mullaney, Thomas S. *The Chinese Typewriter: A History.* Cambridge, MA: MIT Press, 2017.

Nakamura, Lisa. *Digitizing Race: Visual Cultures of the Internet.* Minneapolis: University of Minnesota Press, 2007.

Nakamura, Lisa, and Peter A. Chow-White, eds. *Race after the Internet.* New York: Routledge, 2012.

National Council of Teachers of Mathematics. *Computer Facilities for Mathematics Instruction.* Washington, DC: National Council of Teachers of Mathematics, 1967.

———. *Introduction to an Algorithmic Language (BASIC).* Washington, DC: National Council of Teachers of Mathematics, 1968.

Nelson, Kathryn L. "An Evaluation of the Southern Minnesota School Computer Project, Presented in Foundations of Secondary Education 6353." Master's thesis, Mankato (Minnesota) State College, 1971.

Nelson, Theodor H. *Computer Lib: You Can and Must Understand Computers Now / Dream Machines: New Freedoms through Computer Screens—a Minority Report.* Chicago: Nelson, 1974.

Nielsen, Norman R. "The Merit of Regional Computing Networks." *Communications of the ACM* 14, no. 5 (May 1971): 319–326.

Noble, Safiya U. *Algorithms of Oppression: How Search Engines Reinforce Racism.* New York: New York University Press, 2018.

Norberg, Arthur L., Judy E. O'Neill, and Kerry J. Freedman. *Transforming Computer Technology: Information Processing for the Pentagon, 1962–1986.* Baltimore: Johns Hopkins University Press, 1996.

Oldenziel, Ruth. *Making Technology Masculine: Men, Women and Modern Machines in America, 1870–1945.* Amsterdam: Amsterdam University Press, 1999.

Oldfield, Homer. *King of the Seven Dwarfs: General Electric's Ambiguous Challenge to the Computer Industry.* Los Alamitos, CA: IEEE Computer Society Press, 1996.

O'Mara, Margaret Pugh. *Cities of Knowledge: Cold War Science and the Search for the Next Silicon Valley.* Princeton, NJ: Princeton University Press, 2005.

O'Neill, Judy Elizabeth. "The Evolution of Interactive Computing through Time-Sharing and Networking." PhD diss., University of Minnesota, 1992.

"*Oregon Trail*, Inducted 2016." The Strong, National Museum of Play, World Video Game Hall of Fame. http://www.worldvideogamehalloffame.org/games/oregon-trail.

Oudshoorn, Nelly, and Trevor Pinch. *How Users Matter: The Co-construction of Users and Technologies*. Cambridge, MA: MIT Press, 2003.

Parker, Louis T., Jr., and Joseph R. Denk. "A Network Model for Delivering Computer Power and Curriculum Enhancement for Higher Education: The North Carolina Educational Computer Service." *EDUCOM: Bulletin of the Interuniversity Communications Council* 9, no. 1 (Spring 1974): 24–30.

Parkhill, Douglas F. *The Challenge of the Computer Utility*. Reading, MA: Addison-Wesley, 1966.

Pepple, Steve. "Why Did All Children of a Certain Age Play Oregon Trail?" *Medium*, November 30, 2016. https://medium.com/@stevepepple/why-did-all-children-of-a-certain-age-play-oregon-trail-6a53e27e83d8.

Peters, Benjamin. *How Not to Network a Nation: The Uneasy History of the Soviet Internet*. Cambridge, MA: MIT Press, 2016.

Phillips, Christopher J. *The New Math: A Political History*. Chicago: University of Chicago Press, 2015.

"Pillsbury Co. Purchases GE Computer." *Schenectady (NY) Gazette*, May 1, 1965, Saturday edition.

"Pillsbury Forms Unit to Offer Wide Range of Computer Services." *Wall Street Journal*, January 16, 1968.

Platt, Bill. "Forty Years On: The Changing Face of Dartmouth." *Dartmouth News*, June 17, 2013. https://news.dartmouth.edu/news/2013/06/forty-years-changing-face-dartmouth.

Price, Janet. ["Autobiography."] *The Lathrop Nor'Easter*, series II: vol. 1–4 (Summer 2016): 11. http://lathrop.kendal.org/wp-content/uploads/sites/3/2017/04/NorEasterSummer2016.pdf.

———. "The Use of Conjunctive and Disjunctive Models in Impression Formation." PhD diss., Dartmouth College, Hanover, NH, 1971.

Rankin, Joy Lisi. "Eugenics: Policing Everything." *Lady Science* 33 (June 2017). https://www.ladyscience.com/archive/sciencegenderfascismp2.

———. "From the Mainframes to the Masses: A Participatory Computing Movement in Minnesota Education." *Information and Culture: A Journal of History* 50, no. 2 (2015): 197–216.

———. "Queens of Code." *Lady Science* 9 (June 2015). https://www.ladyscience.com/queens-of-code.

———. "Toward a History of Social Computing: Children, Classrooms, Campuses, and Communities." *IEEE Annals of the History of Computing* 36, no. 2 (2014): 88, 86–87.

Ransby, Barbara. *Ella Baker and the Black Freedom Movement: A Radical Democratic Vision*. Chapel Hill: University of North Carolina Press, 2003.

Rawitsch, Don G. "Implanting the Computer in the Classroom: Minnesota's Successful Statewide Program." *Phi Delta Kappan* 62, no. 6 (1981): 453–454.

Raymond, Jack. "President Asks 1.3 Billion for Missiles, Air Defense as Congress Reconvenes." *New York Times*, January 8, 1958.

Redmond, Kent C. *From Whirlwind to MITRE: The R&D Story of the SAGE Air Defense Computer.* Cambridge, MA: MIT Press, 2000.

Rheingold, Howard. *Tools for Thought: The People and Ideas behind the Next Computer Revolution.* New York: Simon & Schuster, 1985.

Rignall, Jaz. "A Pioneering Game's Journey: The History of *Oregon Trail.*" *US Gamer,* April 19, 2017. http://www.usgamer.net/articles/the-oral-history-of-oregon-trail.

Roberts, H. Edward, and William Yates. "ALTAIR 8800: The Most Powerful Minicomputer Project Ever Presented—Can Be Built for Under $400." *Popular Electronics* 7, no. 1 (January 1975): 33–38.

Rome, Adam. *The Genius of Earth Day: How a 1970 Teach-In Unexpectedly Made the First Green Generation.* New York: Hill and Wang, 2013.

Rosegrant, Susan, and David Lampe. *Route 128: Lessons from Boston's High-Tech Community.* New York: Basic Books, 1992.

Rothman, Josh. "The Origins of 'The Oregon Trail.'" *Boston Globe,* March 21, 2011. http://archive.boston.com/bostonglobe/ideas/brainiac/2011/03/the_origins_of_2.html.

Rudolph, John L. *Scientists in the Classroom: The Cold War Reconstruction of American Science Education.* New York: Palgrave, 2002.

Ruggiero, Vincenzo, and Nicola Montagna, eds. *Social Movements: A Reader.* London: Routledge, 2008.

Sabin, Paul. *The Bet: Paul Ehrlich, Julian Simon, and Our Gamble over Earth's Future.* New Haven, CT: Yale University Press, 2013.

Sandberg-Diment, Erick. "Personal Computers; Musical Software: Point and Counterpoint." *New York Times,* April 12, 1983. http://www.nytimes.com/1983/04/12/science/personal-computers-musical-softwarepoint-and-counterpoint.html.

Saxenian, AnnaLee. *Regional Advantage: Culture and Competition in Silicon Valley and Route 128.* Cambridge, MA: Harvard University Press, 1994.

Schafer, Valérie, and Benjamin Thierry. *Connecting Women: Women, Gender and ICT in Europe in the Nineteenth and Twentieth Century.* Cham: Springer, 2015.

Schein, Edgar H. *DEC Is Dead, Long Live DEC: The Lasting Legacy of Digital Equipment Corporation.* San Francisco: Berrett-Koehler, 2003.

Schulte, Stephanie Ricker. *Cached: Decoding the Internet in Global Popular Culture.* New York: New York University Press, 2013.

Scimeca, Dennis. "Why *Oregon Trail* Still Matters." *The Kernel,* week of August 17, 2014. http://kernelmag.dailydot.com/issue-sections/features-issue-sections/10021/the-lasting-legacy-of-oregon-trail.

Shapiro, Richard. "'World Plan' Week Seeks to Promote Meditation." *Daily Illini,* January 30, 1974, 16, 18. Illinois Digital Newspaper Collections. Archived at https://perma.cc/CE2F-K6VT.

Sharma, Dinesh C. *The Outsourcer: The Story of India's IT Revolution.* Cambridge, MA: MIT Press, 2015.

Sherwood, Bruce. "Status of PLATO IV." *SIGCUE Outlook* 6, no. 3 (June 1972): 3–6.

Smith, Douglas K., and Robert C. Alexander. *Fumbling the Future: How Xerox Invented, and Then Ignored, the First Personal Computer.* New York: W. Morrow, 1988.
Smith, Robert E., and Dora E. Johnson. *FORTRAN Autotester.* New York: John Wiley & Sons, 1962.
Solomon, Cynthia. *Computer Environments for Children: A Reflection on Theories of Learning and Education.* Cambridge, MA: MIT Press, 1988.
Special Correspondent. "Computers in the Home." *Nature* 212, no. 5058 (October 1966): 115–116.
Stern, Alexandra Minna. *Eugenic Nation: Faults and Frontiers of Better Breeding in Modern America.* Berkeley: University of California Press, 2015.
"The Story of *Oregon Trail.*" *Weird H!story Podcast*, March 10, 2016. http://www.interestingtimespodcast.com/2016/71-live-at-the-jack-london-the-story-of-oregon-trail.
Swaine, Michael, and Paul Freiberger. *Fire in the Valley: The Making of the Personal Computer.* New York: McGraw-Hill, 2000.
Taviss, Irene, ed. *The Computer Impact.* Englewood Cliffs, NJ: Prentice Hall, 1970.
Taylor, Ross. "Learning with Computers: A Test Case." *Design Quarterly* 90/91 (1974): 35–36.
Thompson, Keith K., and Ann Waterhouse. "The Time-Shared Computer: A Teaching Tool." *NASSP Bulletin* 54, no. 343 (February 1, 1970): 91–98. doi:10.1177/019263657005434312.
Tilly, Charles. *Social Movements, 1768–2004.* Boulder, CO: Paradigm, 2004.
Turner, Fred. *From Counterculture to Cyberculture: Stewart Brand, the Whole Earth Network, and the Rise of Digital Utopianism.* Chicago: University of Chicago Press, 2006.
Umpleby, Stuart. "Citizen Sampling Simulations: A Method for Involving the Public in Social Planning." *Policy Sciences* 1 (1970): 361–375.
———. "Overstatement." *Change* 6, no. 6 (July 1, 1974): 4.
United States President's Science Advisory Committee, Panel on Computers in Education (John R. Pierce et al.). *Computers in Higher Education: Report of the President's Science Advisory Committee.* Washington, DC: The White House, 1967.
University Communications. "Alma Mater Inducts Bitzer." *NC State News*, March 23, 2011. https://news.ncsu.edu/2011/03/alma-mater-inducts-bitzer/.
Urban, Wayne J., and Jennings L. Wagoner Jr. *American Education: A History.* 4th ed. New York: Routledge, 2009.
Van Meer, Elisabeth. "PLATO: From Computer-Based Education to Corporate Social Responsibility." *Iterations: An Interdisciplinary Journal of Software History*, November 5, 2003. http://www.cbi.umn.edu/iterations/vanmeer.pdf.
Wade, Patrick. "Whatever Happened to: Streaking at the University of Illinois." *News-Gazette*, January 26, 2014. Archived at https://perma.cc/75LT-WG7M.

Walden, David, and Tom Van Vleck, eds. *The Compatible Time Sharing System (1961–1973).* 50th Anniversary Commemorative Overview. Washington, DC: IEEE Computer Society, 2011.

Waldrop, M. Mitchell. *The Dream Machine: J. C. R. Licklider and the Revolution That Made Computing Personal.* New York: Penguin Books, 2002.

Weber, Larry F. "History of the Plasma Panel Display." *IEEE Transactions on Plasma Science* 34, no. 2 (2006): 268–278.

Weeg, Gerard P., and Charles R. Shomper. "The Iowa Regional Computer Network." *EDUCOM: Bulletin of the Interuniversity Communications Council* 9, no. 1 (Spring 1974): 14–15.

Wiener, Norbert. *Cybernetics: Or, Control and Communication in the Animal and the Machine.* New York: John Wiley & Sons, 1948.

Wise, Thomas A. "IBM's $5,000,000,000 Gamble." *Fortune*, September 1966, 118.

———. "The Rocky Road to the Market Place." *Fortune*, October 1966, 138.

Wolfe, Audra J. *Competing with the Soviets: Science, Technology, and the State in Cold War America.* Baltimore: Johns Hopkins University Press, 2013.

———. "Speaking for Nature and Nation: Biologists as Public Intellectuals in Cold War Culture." PhD diss., University of Pennsylvania, 2002.

Wong, Kevin. "The Forgotten History of 'The Oregon Trail,' as Told by Its Creators." *Motherboard*, February 15, 2017. https://motherboard.vice.com/en_us/article/qkx8vw/the-forgotten-history-of-the-oregon-trail-as-told-by-its-creators.

Yates, JoAnne. *Control through Communication: The Rise of System in American Management.* Baltimore: Johns Hopkins University Press, 1989.

———. *Structuring the Information Age: Life Insurance and Technology in the Twentieth Century.* Baltimore: Johns Hopkins University Press, 2005.

Zinn, Howard. *A People's History of the United States.* 35th anniversary edition, with a new introduction by Anthony Arnove. New York: Harper, 2015.

Zweigenhaft, Richard L., and G. William Domhoff. *Blacks in the White Elite: Will the Progress Continue?* Lanham, MD: Rowman & Littlefield, 2003.

Acknowledgments

I ALWAYS MAKE TIME to linger over acknowledgments. I enjoy mentally mapping the networks of collegiality, scholarship, and support that sustain each author. Plus, the acknowledgments sometimes contain a nugget that makes the author or her book especially memorable. And, like most academic books, they follow a formula. You know what they say about rules.

By the time this book is published, my partner, Scott, and I will have spent nearly half our lives together. He encouraged me to pursue my doctorate, and he rode the ensuing rollercoaster of academic life with me. He listened to my first thoughts about archival finds, and he asked perceptive questions. He heard conference talks and job talks, and he read—and later proofread—draft upon draft. Scott loves cooking, and I have been nourished by his countless scrumptious meals and innumerable acts of encouragement and love—not to mention his awesome partnership in parenting.

Our daughter, Lucy, has grown up with this book. I conducted significant archival research when I was pregnant with her. I started writing after she was born, and she gained years, feet, and inches while the book gained chapters. The anticipation of family time with Scott and Lucy—building Legos, bicycling, "reading breakfasts"—honed my focus when I wrote. As I pen (or, more accurately, type) this a few weeks before her sixth birthday, she knows that I am immeasurably proud of her.

I extend my heartfelt gratitude and admiration to my writing group: Mary Brazelton, Amy Johnson, David Singerman, and Emily Wanderer. We have shared drafts, deadlines, laughter, and lamentations for over five years. Their questions and suggestions vastly improved this book, and their advice and friendship buoyed me during the dark times. Here's to another five years (at least), Boston Writing Group!

While writing, I also benefited from the collective knowledge and hospitality of three wonderful academic communities: Yale University's Program in the History of Science and Medicine; the MIT Program in History, Anthropology, Science, Technology, and Society; and the American Academy of Arts and Sciences. I am especially grateful for the generosity of my mentors, Daniel Kevles and Glenda Gilmore at Yale and David Mindell at MIT; they have been my academic champions. At Yale, Paola Bertucci, Joanne Meyerowitz, Naomi Rogers, Paul Sabin, Bruno Strasser (now at the University of Geneva), John Warner, and Bill Rankin honed my thinking and my writing, and cheered me along the way. (For readers interested in connections and coincidences, Bill and I are not related, nor am I related to the historian Alisha Rankin at Tufts.) At MIT, Harriet Ritvo, David Kaiser, Heather Paxson, Stefan Helmreich, and Hanna Rose Shell nudged me to the intersection of history and science and technology studies, and thereby strengthened this book. I owe special thanks to my Yale colleagues Helen Curry, Robin Scheffler, Cecilia Cárdenas-Navia, Rachel Rothschild, Justin Barr, and Gerardo Con Diaz, and to my MIT colleagues Steven Fino, Rebecca Woods, Julia Fleischhack, Renée Blackburn, Mitali Thakor, Shreeharsh Kelkar, and Ellan Spero for their collegiality and encouragement. At the American Academy, Larry Buell offered steadfast guidance and the inspiration for the title of this book, and my fellow visiting scholars shared meals, books, and insightful cross-disciplinary analysis. Thank you, Katherine Marino, Emily Owens, Lukas Rieppel, Merve Emre, Rachel Guberman, Rachel Wise, and Les Beldo.

The archives in which I spent most of my research time felt like homes away from home, and for that I recognize the warmth and intellectual curiosity of their staffs. At the Charles Babbage Institute (CBI), I especially benefited from conversations with Tom Misa, Jeff Yost, Arvid Nelsen, Amanda Wick, and Susan Hoffman. Moreover, Susan (at CBI) and Elisabeth Kaplan at the University of Minnesota Archives worked together on my behalf to ensure that

I had access to every possible collection of interest. At Dartmouth College, I could not have navigated all the relevant special collections without the guidance of Sarah Hartwell, Peter Carini, Morgan Swan, Barbara Krieger, and Phyllis Gilbert. Jill Vuchetich facilitated my visit to the Walker Center Archives, and Hampton Smith helped me work through the bountiful boxes at the Minnesota Historical Society. Lisa Renee Kemplin and Jameatris Rimkus at the University of Illinois Archives fielded questions and found photographs. Lindsay Anderberg at the Bern Dibner Library, Massimo Petrozzi at the Computer History Museum, Tim Noakes at Stanford Special Collections, Karen Parsons at Loomis Chaffee, and Paige Roberts at Phillips Andover located precious materials and photographs.

Writing these acknowledgments has brought me the joy of realizing how much I have relished the writing process, especially sharing my work at conferences. I am grateful for the stimulating commentary, observations, and questions I received at the following venues: the History of Science Society, the Society for the History of Technology, the Special Interest Group for Computers, Information, and Society, the MIT Symposium on Gender + Technology, the Charles Babbage Institute, the Yale Digital Humanities Working Group, the Yale Environmental History Colloquium, the STGlobal Consortium, the American Society for Environmental History, the Society for Cinema and Media Studies, and the New York University Department of Media, Culture, and Communication.

I offer my profound thanks to the IEEE, whose Fellowship in The History of Electrical and Computing Technology enabled me to conduct archival research and to begin the writing process. The Adelle and Erwin Tomash Fellowship, awarded by the Charles Babbage Institute, further supported my research and writing. Yale University generously awarded me various fellowships, and its Program in the History of Science and Medicine supported both conference and research travel.

Chapter 5 is a revised and expanded version of "From the Mainframes to the Masses: A Participatory Computing Movement in Minnesota Education," which I originally published in *Information & Culture: A Journal of History* (Volume 50, Number 2, pp. 197–216.

Copyright © 2015 by the University of Texas Press). I thank the University of Texas Press for permission to republish material from that piece here.

Sharmila Sen, my editor at Harvard University Press, understood and advocated for my vision for this book from the first minute of our first conversation, and Amanda Claybaugh at Harvard University connected me with Sharmila in the first place. They both have my eternal thanks. Heather Hughes at the Press provided invaluable assistance, allegiance, and enthusiasm for this book, in addition to our shared genealogy in the *Oregon Trail* generation. I greatly appreciate the consideration demonstrated by the two anonymous readers, and I hope they see their helpful suggestions reflected herein. Audra Wolfe (at the Outside Reader), Julia Kirby (at the Press), and Liz Schueler and John Donohue (at Westchester Publishing Services) provided meticulous editorial review.

For standing by me in 2017, thank you, Katie Herrold, Leila McNeill, Anna Reser, Kate Sheppard, Meryl Alper, Mar Hicks, Stephanie Dick, Emily Richmond Pollock, Miriam Sweeney, Kevin Driscoll, Nabeel Siddiqui, Nathan Ensmenger, and Andy Russell. For reminding me to breathe and move, to be grounded and to fly, thank you to the yoga communities of Black Crow in Arlington, Massachusetts, and Hilltop in Lansing, Michigan. For a crucial conversation at just the right time, and for the inspiration of her scholarship, thank you, Deb Harkness. Michael Lisi, thank you for being my oldest friend, cheering for me and laughing with me since before my fifth grade spelling bee. For encouraging my love of learning, thank you, Carol and Perry Lisi. And once again, from beginning to end, cover to cover—for everything—thank you, Lucy and Scott.

Index